BARN BUILDING

THE GOLDEN AGE OF BARN CONSTRUCTION

BARN BUILDING

THE GOLDEN AGE OF BARN CONSTRUCTION

Jon Radojkovic

A BOSTON MILLS PRESS BOOK

Published by Boston Mills Press, 2007
132 Main Street, Erin, Ontario N0B 1T0
Tel: 519-833-2407 Fax: 519-833-2195

In Canada:
Distributed by Firefly Books Ltd.
66 Leek Crescent
Richmond Hill, Ontario, Canada L4B 1H1

In the United States:
Distributed by Firefly Books (U.S.) Inc.
P.O. Box 1338, Ellicott Station
Buffalo, New York 14205

The publisher gratefully acknowledges for the financial support of our publishing program, the Canada Council, the Ontario Arts Council, and the Government of Canada through the Book Publishing Industry Development Program (BPIDP).

Library and Archives Canada Cataloguing in Publication

Radojkovic, Jon, 1952–
Barn building : the golden age of barn construction / Jon Radojkovic.

Includes bibliographical references and index.

ISBN-13: 978-1-55046-470-2
ISBN-10: 1-55046-470-1

1. Barns — United States — Design and construction — History. 2. Barns — Ontario — Design and construction — History. 3. Barns — Québec (Province) — Design and construction — History. 4. Barns — United States — History — Pictorial works. 5. Barns — Ontario — History — Pictorial works. 6. Barns — Québec (Province) — History — Pictorial works. I. Title.

NA8230.R325 2007 728'.9220973022 C2007-901750-9

Publisher Cataloging-in-Publication Data (U.S.)

Radojkovic, Jon, 1952–
Barn building : the golden age of barn construction / Jon Radojkovic.

[192] p. : photos. (chiefly col.) ; cm.
Includes bibliographical references and index.

Summary: A photographic celebration of the era of North American barn-building known as "The Golden Age of Barn Construction," from the early 1700s to the mid-20th century.

ISBN-13: 978-1-55046-470-2
ISBN-10: 1-55046-470-1

1. Barns — United States — Design and construction — History. 2. Barns--Canada — Design and construction — History. 3. Barns — United States — History — Pictorial works. 4. Barns — Canada — History — Pictorial works. I. Title.
728/.9220973 dc22 NA8230.R336 2007

Edited by Anne Judd
Design by Gillian Stead
Illustrations by Mary MacCarl

All photographs by Jon Radojkovic unless otherwise noted.

Printed in China

Acknowledgments

First of all I would like to thank my wife and partner, Lillian Burgess, for her many helpful and thoughtful suggestions and time spent looking over my work. Thanks to Mary MacCarl for taking my rudimentary drawings and turning them into wonderful pen and ink illustrations.

So many people took the time to give me valuable information or meet me in various places across the northeast of the USA and Canada and show me amazing barns. These included Jan Corey Arnett, John Holcombe, Jack and Anne Lazor, Andrew Levin and Mary Ormrod, Francis Loomis, Jack Worthington, Alan Seigworth, Frances Walbridge and Willis Wood. A special thanks to Tom Denison who has gone out of his way to provide me with books on barns and local history. Jeffrey Alford and Naomi Duguid supplied me with timely advice and support. Arthur Plumpton was generous with valuable photos and historical information. Thanks to Pief Weyman for help reproducing some of my archival photos. Thanks also to managing editor Noel Hudson, manuscript editor Anne Judd, and to publisher John Denison, for believing in my idea.

Barn associations, preservation societies, and museum and library personnel contributed by providing me with photos, locations, and information on historic barns. These include the Barberton Historical Society of Ohio; Friends of the Thumb Octagon Barn, Michigan; The 1000 Islands Land Trust, New York; Friends of the Star Barn Society, Pennsylvania; Wellington County Museum and Archives, Ontario; and my own local library, the Chesley Centennial Library.

To my growing-up and grown-up sons, thanks for being patient with me.

Contents

Bank Barns *continued*

Lovely tall cupolas and dormers grace this beautiful green gambrel-roofed barn in Lena, Ohio, still a working dairy barn.

Introduction

THIS BOOK CELEBRATES BARNS. It explores their variety; pays tribute to the craftsmanship that formed them; acknowledges those who built them and the life they led. It also tells the story of my trips to see every one of these barns, visiting with people I had never met before but with whom I shared a bond. I learned about their lives through the centerpiece of their farms, the barns.

There were many more barns at the end of the nineteenth century when this estate auction took place near Oxford, New York. Our historic barns are disappearing at a fast rate, including the three we see in this photo taken about 1900.
Courtesy of Richard Place

Farming has evolved immensely since Europeans settled this continent in the seventeenth century. We have moved from sickles to grain cradles to horse-drawn reapers to threshers and combines. The last one hundred years especially have seen even greater changes in farming, as technology's rapid advance has transformed agriculture in the USA and Canada — and with it a way of life.

Barn building has kept pace with the times. For pioneers, log barns were the practical buildings of the day. As communities became established, settlers were able to accomplish the craftsmanship of timber-frame Dutch and English barns and long, low Quebec-style barns. Such designs came from ethnicity, the country of settlers' origin and the building techniques early pioneers brought. Dutch and Swiss settlers built boxy structures with huge timbers available from the virgin forests in New York State. The English brought the tall, elegant, post and beam structures that were practical to build and adapted well to the New World climate. The French introduced a form that came with building specifications written down by the governor of New France.

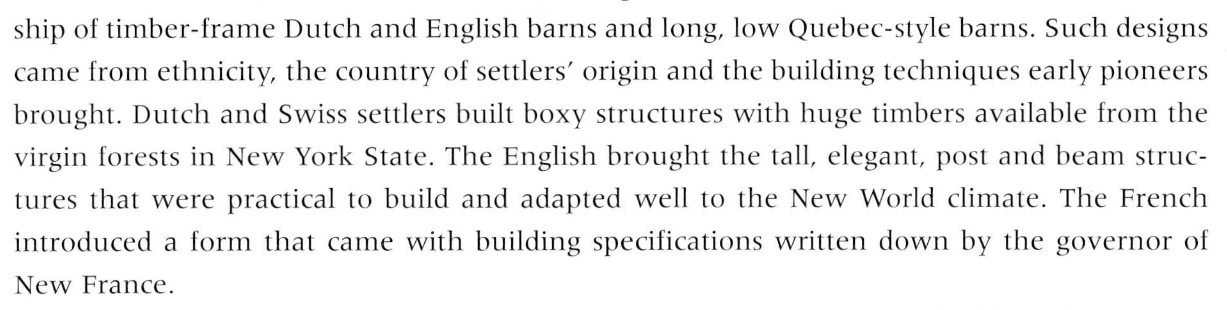

Although ethnicity played a large part in early barn construction, builders also adapted their techniques to North American weather. After all, it was our extreme climate that demanded barns. The colder the winters, the bigger the barns. By the beginning of the 1800s

Vermont farms were settled 150 years ago wherever there was a bit of good land. Today tourists flock to this mountainous state, especially in autumn, to enjoy just such sights as this near Cambridge — a moment's flash of red barn board, a cupola, a neat roof, the brilliant leaves and then it's gone.

barns were recognizably New World in nature, adapting to local conditions and using local materials. Dutch barn designs evolved using smaller timbers; the English barn was expanded and stables added under its frame. The concept for bank barns originally arrived with the Germans, but the buildings soon grew larger and wider, with ramps at the gable ends as well.

During this period life on farms pretty well carried on in its traditional form, with members of the family having their own specific tasks. Men built barns, cared for animals and tended fields, fixed fences, cleaned out stables, pruned orchards and maintained farm equipment. Women raised children, prepared food, washed laundry, prepared herbal medicines, sewed clothes, made soap, fed chickens, churned butter and cooked feasts for barn raisings. Children as young as three or four gathered berries and nuts, weeded gardens, picked stones from fields and helped mother with washing and peeling vegetables. The whole family came together for butchering a pig or a cow.

On an even a bigger scale, communities helped one another in a number of ways, from "bees" for quilting to threshing to raising barn structures. If a neighbor needed a hand, help was on the way. Socially, neighbors would see each other regularly, eat together often, go to church and attend each other's weddings and funerals. So often I have heard from the older folks I talked with, "We couldn't have done it without the help of neighbors."

That rural social fabric has been frayed, but strands remain in the historic farms still dotted across the USA and Canada. It's that heritage that I wanted to celebrate and learn from.

My barn discovery travels have taken me to Michigan, Ohio, Pennsylvania, New York, Vermont, New Hampshire, Ontario, and Quebec. The many miles I have driven have rewarded

The incredible gold of a field of mustard, now called canola, produces valuable oil, making it an important crop in the Chesley, Ontario, region. Built in the 1880s, the barn is a standard size for this area, 40 feet by 60 feet.

me with an untold number of exhilarating and unforgettable experiences. I have visited and photographed every one of the barns in this book and was fortunate to enter many of them. No two barns I have seen were alike. Even those built by the same builder were fine-tuned to the owner's specifications, the lay of the land and type of farming operation. Barns were "handmade," not built inside a factory mass-producing trusses and walls that today do make many barns exactly the same.

We define heritage as an inheritance from the people before us which can be learned from and built upon. Many farm families value their old barns as their link with the past. Barns help us perceive why we are what we are today. My travels have certainly made me understand even more the richness of our landscape in rural history. Notching timbers is a skill that has been handed down through the generations but I have been lucky enough to learn it by teaching myself and from books. I learned to imagine the structure, then to put that concept onto wood joinery. It takes patience and perseverance, richly rewarded in having a mortise and tenon fit precisely.

A narrow, kit-built barn near Meredith, Michigan. Farmers could order barn kits from many suppliers, even Sears, Roebuck and Company, and have them delivered to the nearest train station. Or they could build from a book of barn plans. Barns like this with its curved roof and white and red painted doors are often seen in Michigan's countryside.

Before notching a timber frame, I start by ordering the timbers at the local Amish sawmill. Since they don't have a phone, it means a two-mile trip. Often teenagers and young men work the demanding jobs at these sawmills. Dressed in their traditional homemade dark blue jackets and pants, often ripped and repaired, with either straw or black, wide-brimmed hats, they first talk with me about the weather and local gossip. They love to laugh and so humour is much a part of our conversation. Then, with my timber list in hand, we'll head out toward the log yard, to choose from the great piles of pine or white ash logs large enough to make the beams I need. We check them for soundness and make sure they are straight.

A few weeks later, I'll drive over with my trusty (if somewhat rusty) 1979 Dodge pick-up truck and a sturdy wagon. Green-cut timbers are heavy, and it takes a big heave-ho to load them all on the wagon and back of the truck. Sometimes I take two trips, depending on the size of the order. I stack and let them dry for a few weeks, in the meantime drawing up the cutting list, deciding on the lengths and what notches I will need to cut.

Traditionally, timber framers used augers or boring machines to drill out the holes for the mortises, the hardest joint to notch, since they involve removing solid wood leaving a square hole. I used a boring machine to notch my house frame before we had electricity here but, when the "wire" came, I began using electric drills. The final cleaning out of the notch I still do by hand chiseling. It takes patience, experience, and physical endurance. I love the feel of a large, sharp chisel that can take off a sixteenth of an inch or one inch of wood, always with the pounding of the mallet, the flying chips of wood. Although notching is a lot of physical work, it challenges my mind too. I twirl the structure around in my head, making sure I am notching correctly. A single timber may have many notches and one mistake would mean days of work put aside for the stove. Nothing feels better in the world for me than to see a pile of timbers, all notched, waiting to be put together as a house or barn.

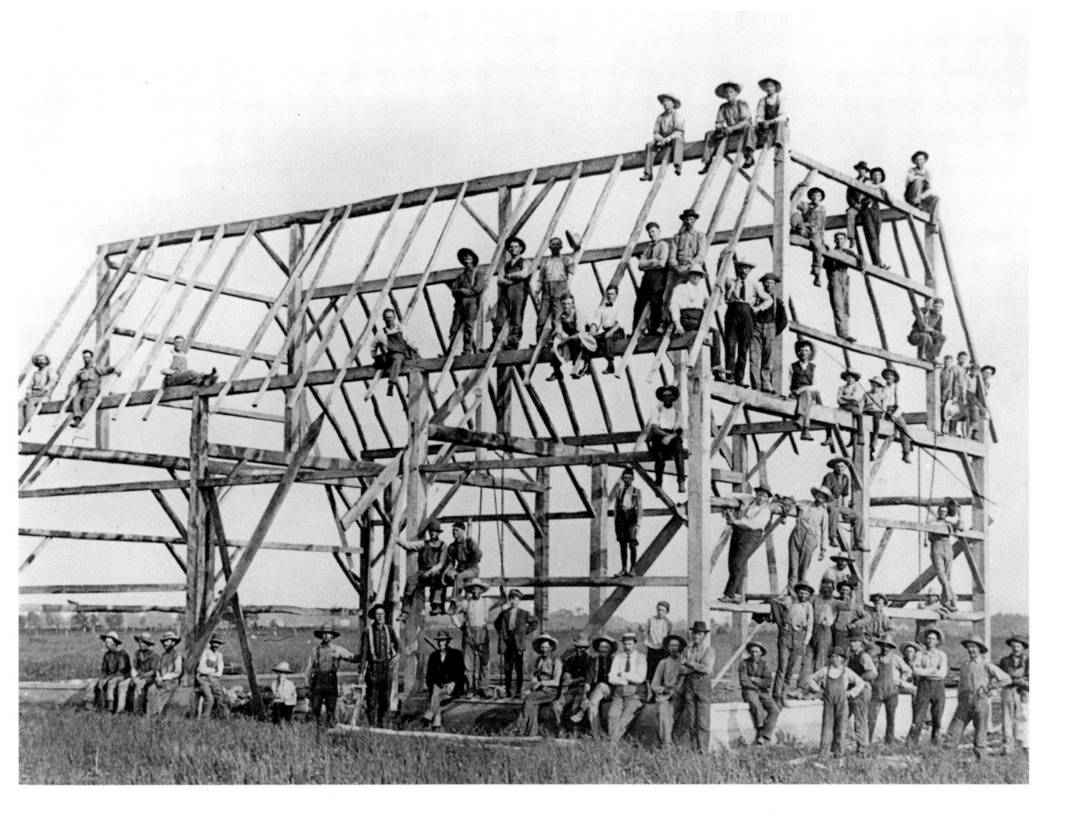

This late nineteenth-century barn-raising photo shows the typical tall, narrow, gambrel-roofed Michigan barns built during that era. (See page 176)
Courtesy of State Archives of Michigan.

Raising a timber frame still excites me. Although a hundred years have passed since barn raisings were common, they have a peculiar and enduring fascination for us. They carry the aura of a past, of strong community bonds. Raisings brought the whole neighborhood together. Besides the practicality of putting up a barn frame for a local farmer, the raisings also forged new relationships — friendships and sometimes life-time marriages.

I find something incredibly moving when a pile of squared timbers lying inert in the morning has been transformed by nightfall into a graceful yet immensely strong and durable frame held together with nothing but pegs of wood. Perhaps the grassroots human scale of it all touches everyone who participates in a raising.

Three of the dozen sandstone pillars that support a corn-drying shed on an abandoned farm near Summerville, Pennsylvania.

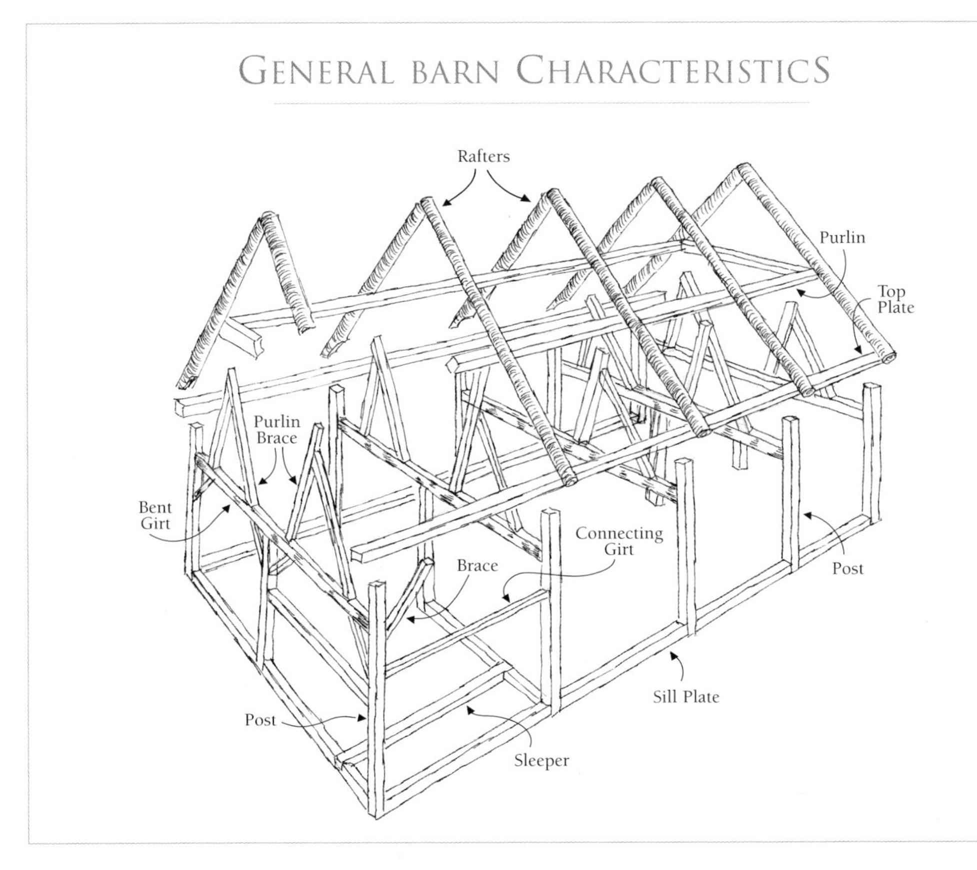

Of course I do acknowledge that many old barns are falling down from neglect, not fitting into current agricultural uses, or not being needed by non-farming owners. Fortunately, preservation associations have formed to save these barn structures. Join one today, because every year, hundreds of barns are taken down, the good timbers to be made into flooring, the bad pieces burned and the occasional lucky few structures to be raised again as part of another barn or house.

The loss of barns is certainly one of the reasons I wrote this book, for all of us to remember an amazing era of barn building from the early 1700s to the middle of the twentieth century. I like to call it "The Golden Age of Barn Construction".

A typical English three-bay post and beam barn, the Anderson barn was built during the 1880s, near Peabody, Ontario.

Standing at the end of a quiet road next to the Madawaska River, near Latchford Bridge, Ontario, this log barn is still used for storage. The bottom long side was last chinked with a lime/cement mixture, which has mostly fallen out. The top half, left unchinked, allowed air movement in the barn to help keep the hay dry.

Log Barns

Introduction

LOG BARNS AND OUTBUILDINGS have an integral connection to the pioneers who came to eastern North America, where they cleared land, grew crops, raised livestock and built houses and barns to keep themselves and their animals. The few log buildings that remain show us tangible evidence of that era.

In my many trips across eastern USA and Canada I occasionally passed by a log barn and always felt an emotional pull that is hard to define. The difference between our world today and that of the pioneers who built those log structures is so great. Imagine cutting down huge trees with a bucksaw, making a small sunlit clearing for a home, with the nearest neighbor being perhaps miles away, everyone caught with the same fever — a fresh start, a new life!

Log barns were generally the first permanent structures built by settlers as they came inland. There was an abundance of trees and the methods were simple and adaptable. As the new families traveled to their new lands, they passed many examples of log buildings put up by previous settlers who had brought some knowledge from Europe. Newcomers absorbed techniques which they soon put to use.

At first a small temporary lean-to made of poles with bark for roofing housed the family.

All that remains of the village of Musclow, Ontario, settled in the 1860s by the Musclow family, are a few barns and a couple of houses. Gone are the stores, the post office, the gristmill and one-room schoolhouse.

An old-fashioned farm scene near Eganville, Ontario: log barns, horses, and all-wood fencing. The main barn still serves a purpose, storing hay and sheltering animals. Chinking was applied only on the bottom half of the barn to keep the outside air from blowing on the animals inside during the winter, while the top loft half is not chinked to let in air to keep the hay dry. Note the joist ends sticking out of the barn, marking the second loft area and the dovetail joinery at the corners of the barn.

Then, often before a cabin, they constructed the first barn, made of logs cut and fitted with the help of neighbors. This building stored their vital crops and livestock: usually an ox, a pig, and later a milking cow and some chickens, all essential for survival.

A hundred or two hundred years later, we see these log barns and outbuildings in a clump of trees, usually surrounded by large cropped fields, mostly forgotten, or used as storage sheds. We easily forget ordinary people made them with sweat and urgency.

This style of building dates back almost 12,000 years to northern Europe. Swedes and Finns brought log building techniques to Delaware during the 1630s. Germans introduced them in the early 1700s in Pennsylvania.

Log barns were an "every man's barn." If you could build a box then you could build log structures. The simplest log building is a rectangular box made of horizontally laid logs notched

Another little side road in this rocky overgrown north country and more log barns unfortunately no longer used for farming. The long grass and texture of the logs gave me a feeling of times gone by, of a once-vibrant farming community united in making a living from the land. A good example of barn expansion — just add a few logs, a roof, and presto! more barn space. Near Killaloe, Ontario.

This example of a full dovetail notch shows the intricate ax work and the chinking between logs.

together at the corners, a basic form known as a single crib. Put a roof on a crib and you have a barn or house or whatever outbuilding you need. A double crib log structure was commonly used to build larger barns. Its two boxes or cribs of logs stand side by side, with a space between covered by a common overhanging roof. The cribs stored hay, stooks of grain, and livestock, while the breezeway between the two cribs would be used for threshing and for storing implements.

Assembling corners was the main building challenge because that would determine how closely the logs fit together and therefore how much chinking needed to be done. The simplest corner is the saddle notch with projecting round logs. The bottom of each log was hollowed with an adz or ax to fit the log below. The *V* notch is more characteristic of German log building. Each peeled and roughly squared log is hewn with an inverted *V* on the top and on the underside to create a tight fit.

The full dovetail is the most common type of corner joinery that has survived from log barn building. More skill goes into its cuts: a compound angle at both top and bottom faces of the log, making a lasting strong fit. This notch was later simplified and known as the half dovetail. The top edge of the peeled, squared log has a simple notch angled downward from back to front while the bottom notch angles upward from the end.

Another way of placing logs was the French *pièce-sur-pièce* or piece on piece. Essentially a post and beam framework, this method used upright logs at the corners, in-filled with equal-length horizontally laid logs. Each corner and middle post was slotted. Corner posts were slotted on the inward side and middle posts on both sides. The horizontal log ends were chiseled to fit into these slots and then slid down between the two vertical posts. This system had advantages. Logs could be shorter and smaller with a vertical post every six feet. Only two people could lift a log into place without levers or block and tackle. We find *pièce-sur-pièce* construction in Louisiana, parts of Michigan and Maine, and commonly in Quebec.

The windmill played a large part in early farming, pumping water for the house and barn and also powering machinery to grind grain. The nineteenth-century log barn is still well maintained on this farm near Cormac, Ontario.

The Stong double-crib log barn, built in 1825, has log cribs forming its gable ends.

Black Creek Pioneer Village, Ontario

Pennsylvania German settlers came to this area, just north of present day Toronto in the early nineteenth century. Johannes Schmidt built a sawmill in 1808 on Black Creek and soon the community of Edgeley formed around it. Almost 200 years later this little village has been preserved in the form of a living museum called Black Creek Pioneer Village.

There is an excellent example of a double crib log barn, still on the original site in the village, built by Daniel Stong in 1825. The crib logs form the walls on the gable ends while a large common overhanging roof and posts form the long walls. There are intricate dovetail notches for the corners while the breezeway forms the double threshing floors. Inside is a rare Conestoga wagon, such as the ones used by settlers heading north from Pennsylvania. The cribs were used to store grain, mostly wheat for making flour, and hay for the livestock.

Girls dressed in period clothes.

One of the two log cribs inside which also supports the roof.

A rare frame and solid wooden wheels of an early eighteenth-century Conestoga wagon, used for carrying freight and the belongings of pioneers.

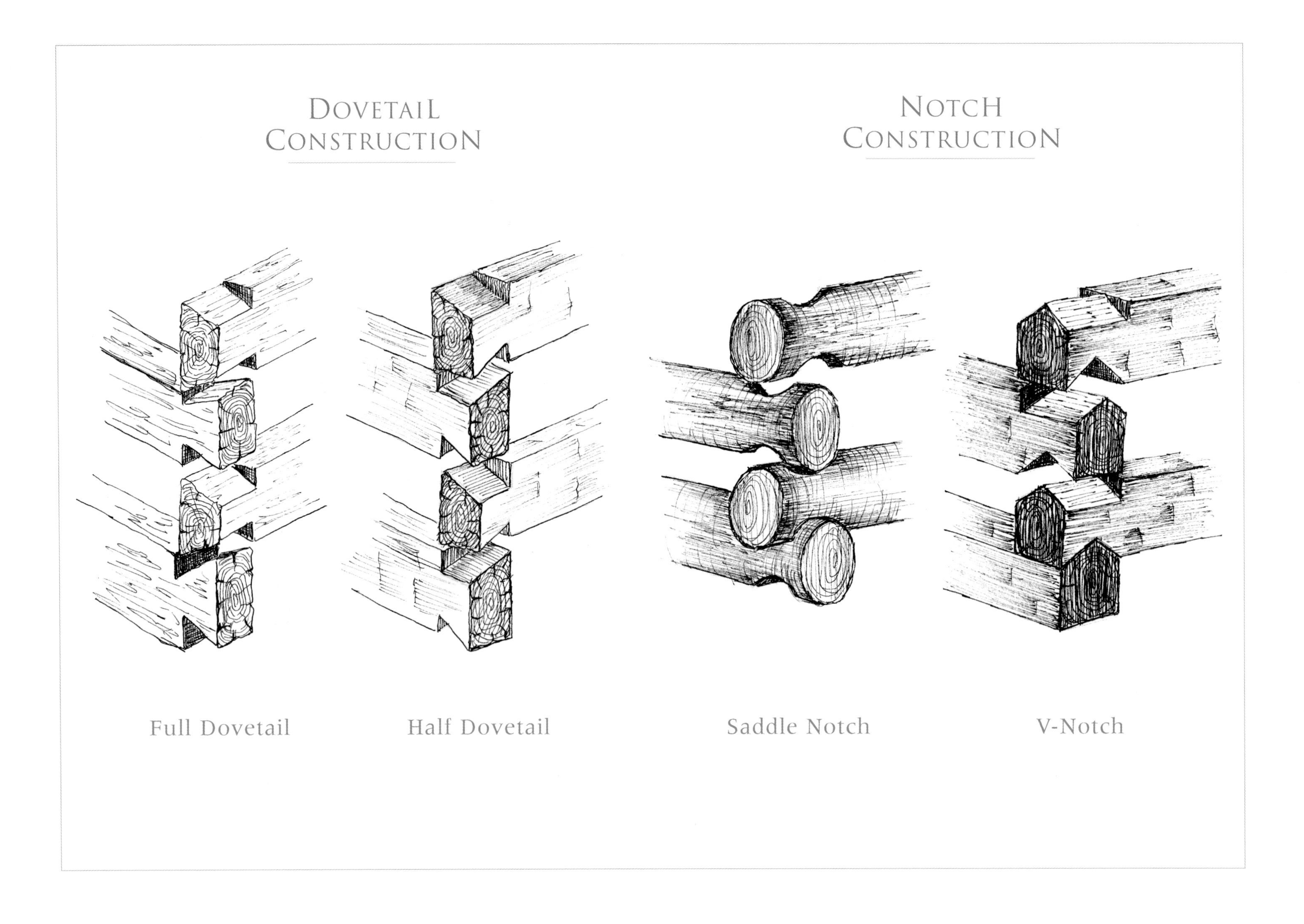
DOVETAIL CONSTRUCTION
NOTCH CONSTRUCTION
Full Dovetail
Half Dovetail
Saddle Notch
V-Notch

With any method, spaces between logs needed filling. For chinking, settlers often used peat moss; mud and straw were also utilized and later lime and mortar. Log barns were typically chinked only half way up the walls, to protect the livestock from the wind, rain, and snow, while allowing air to ventilate the top part.

At first builders also made gable ends from logs, the free building material. Later, sawmills made boards used as horizontal clapboard, or vertical boards nailed to supports inside.

The biggest challenge facing early barn builders was constructing a tight roof. Some of the first roofs were made from hollowed out basswood trunks, laid from the eaves to the ridgepole, much like European clay tiles, one rounded *U* fitting over another upside down one.

One of the three log barns still standing in the historic village of Musclow, Ontario, this small building housed the milking cow and her calves with storage above for hay.

Thatch was often used initially, another method brought from Europe but thatching material native to North America was not as effective. Sometimes pioneers tried a combination of straw with poles laid on top. When they could find straight-grained hardwood, especially ash, they could make shingles with a froe and mallet. After they were split, the shingles were soaked in water and applied wet, to prevent splitting when nailed. Hand-forged square nails would bend in dry hardwood shakes. Found in some areas, cedar made shakes which didn't need to be soaked unless very thick.

Although the earliest forms of log building were brought to North America by Northern Europeans, each with their own specialty, once here combined with other cultures to produce a variety of log buildings now simply known as "local" architecture. The Irish and Scottish brought with them their knowledge of stone masonry, making beautiful stone foundations for later log buildings or additions. Others brought designs with high-pitched roofs, an improvement because they shed rain and snow best. Some settlers in-filled between log posts with mud and rock rubble, or fine stone work.

Log or crib barns are especially numerous in the Appalachian and Ozark Mountain states of North Carolina, Virginia, Kentucky, Tennessee, and Arkansas. The Jones Log Barn, built circa 1730 in Pennsylvania, is one of the oldest log barns in the Mid-Atlantic region. As well, the hilly countryside in eastern Ontario is a wonderful place to see log barns. Hastings County was colonized in the 1860s. Settlers cleared the land but found it very stony and not fertile. Many left for the west in the late nineteenth century. The ones who stayed made a subsistence living, building log barns as needed, but never improving or stepping up to timber frames, as the land never seemed to provide enough capital.

We are fortunate to have these log barns still standing because the land was too poor to provide farmers with a good living. Some are still in use but unfortunately many are quietly decaying into the ground. Still, they are a testament to the strength and longevity of mostly cedar and pine log structures that are more than a century old.

The vast area provided by a barn wall or roof has often been used as an advertising billboard. Companies get good exposure for their products with little effort and farmers get some extra, often sorely needed, cash. However, the juxtaposition between such a primitive-looking structure and modern graphics seemed sad to me, a pointer to the future of neglected barns. This barn is close to Wingle, Ontario.

Amazing to me are the number of log barns one sees on each farmstead. As farms expanded and settlers cleared more land, they needed more space for crops and livestock. Instead of erecting additions to the barns, they constructed new log structures — one for the pigs, one for the cows, the horses, the chickens, grain, hay and so on.

Sometimes log barns were connected in a courtyard style, forming a *U* or a complete square. Others were simply built where they were needed, some in a not-so-straight line, others connected, some standing across the laneway from each other. This progress made sense: cut a few logs during the winter, build a crib in the spring and add a roof. Farmers are a pragmatic lot and during the pioneer era, this type of construction was practical, simple and accessible.

The KARGUS Log Barn

During the 1860s, as European settlers flocked to claim land in the wild country of northeast Ontario, the government of the day gave them a list of provisions needed. When pioneering farmers arrived in this region, there were, of course, no stores, no settlements, nothing but unbroken, forested lands. For a family of five the list included such things as 8 barrels of flour, 80 bushels of potatoes, 1 ax, 2 augers, 3 blankets, 1 pig, 1 teapot, 1 cast iron frying pan, 1 handsaw and 1 window sash for their initial log dwelling.

As settlers slowly began to develop their farms, the first permanent agricultural buildings were log barns. This region is a prime example showing how lack of good fertile soil has kept log barns still in use as farmers could never afford to upgrade. Time stood still here and, because of that, we can see remarkable examples of what early log barn construction would have looked like in most parts of the Great Lakes region.

Farmers in the Ottawa Valley expanded by building more barns of the same size instead of larger ones.

Farmers expanded their operations here by simply building another log barn, and then another and another as needed. On the Kargus farm, five log barns were built before the 1900s. Looking at the barns from the road, the first barn on the left had three livestock doors. The first door accommodated pigs, with a loft for straw and hay and a small enclosure outside also made of logs. The next opening was for calves, the third for the family milk cow.

The next barn, larger, and also attached to the first, housed the backbone of the farm, the horses. Freddy Kauffeldt, whose uncle once owned the Kargus farm told me there were as many as six: four large draft animals

The inside of the largest barn on the Kargus farm once housed hay and straw. The roof has been renovated recently using new trusses.

used for farm work and two smaller and faster standard-bred horses for pulling buggies or sleighs. A large door on this barn allowed the big working horses easy entrance.

A smaller, separate barn for chickens built next to these, had a loft for them to roost during the long, cold winters. At the end of the line of four barns stood the granary, the only post and beam building. This strong construction could withstand the force of many tons of grain pushing against the walls.

Across from the horse barn stand the two largest barns, joined together at the middle. The length of logs available determined the span of the barn. When farmers needed more space, they cut more trees and stretched the building by adding to it. A common joining wall remained with a large door opening between the two structures.

The two largest Kargus barns were used for the storage of hay, straw, and farming equipment. Cattle stayed outside all winter behind this hay barn in an enclosure made of split-cedar rails.

Since the thinly soiled land was quickly depleted, many farm families left after one generation. Those who stayed continued to farm under harsh conditions and found seasonal employment in logging camps and mines in the area.

"After the '50s and '60s there were just less and less people," said Freddy. "You can't make a go of these farms," he added. Now many of these early log barns stand unused, waiting for rot to slowly take hold, or a match to bring a quick end or a family from the city to use them for recreational purposes.

The end of a joist log sticks out above a doorway as well as a hanging downside horseshoe placed there for good luck.

The oldest farm in the Mohawk Valley, New York, is also home to a Dutch-style barn built around 1760. The Mabee colonial Dutch house was built in 1670, with slave quarters attached, and remained in the Mabee family for almost 300 years. This New World Dutch barn was moved to the present site near Rotterdam Junction in 1999, and is now a living museum on the Mabee Farm Historic Site.

Dutch and English Barns

DUTCH- AND ENGLISH-STYLE BARNS represent the two oldest timber-framed techniques of construction brought over from Old-World Europe.

Dutch Barns — Introduction

I was cruising along just north of Albany, New York, after a day spent looking for Dutch barns. I hadn't had much luck until I drove by a cluster of buildings beside the four-lane highway and, set in the middle, I saw a beautiful Dutch barn, in almost original condition. I abruptly slowed and, without signaling, turned off, my head spinning at having found such a prize. The driver behind me had to stop as quickly, honked, and probably muttered something like "damn tourist."

The Dutch Barn has certain defining characteristics: high-pitched roof, low eaves, horizontal siding, bird entrances near the peak and a wagon door at each end of the gable.

I found John Wiggen behind the barn, dressed in overalls and hammering away on another project, a reproduction of a colonial building. The Dutch barn was not a reproduction, but had been meticulously restored by John over the course of many years. What struck me most about John was the effort and pride he put in into restoring and recreating buildings from colonial times. "History is important," he said. "We can learn a lot from it."

Walking into what author John Fitchen calls a *New World Dutch Barn* is like entering a great cathedral. Huge columns of timbers rise forming aisles, crossing with even more massive wood beams to create large open spaces in the middle. With a steep-pitched roof, the inside resembles a place of worship. This is no coincidence, since Dutch barns emulated the many churches and cathedrals in Europe built in this same tradition.

Now used as a storage shed, this unpainted Dutch barn is identifiable by its high peak and square lines. I saw this one just south of Middleburgh, New York, where many of these eighteenth-century barns were built along the Schoharie River.

In the Old World farmers used stick and daub to cover their barn walls, while in the New World, wood siding was used.

Dutch barns have their roots in a history that dates back almost 2,000 years. Until the seventeenth century, the family lived inside the barn. Today these great barns are one of the last reminders of the pre-industrialized heritage of New York and New Jersey.

An estimated 1,600 New World Dutch Barns were built between 1630 and 1825. Those still in existence are the oldest post and beam barns in North America. Typically the framework is three to five bents long but usually wider than long, giving a boxy appearance. Unpretentious-looking from the outside, Dutch barns have wagon doors in both gable ends, a cattle door in one gable end as well and bird holes, found in many different shapes, at the peaks. One noticeable difference from most other barns is that they normally have no windows. Because of the extra width, the eaves are low, and in Old Europe, where the Dutch, Germans, and English built these aisled barns, the roof came down almost to the ground. In the Old World the walls were stick and daub (a mixture of earth, straw or grass, and sometimes lime or cow dung) while early North American Dutch barns had wide horizontal siding which shed the rain and snow effectively.

Inside, the fine craftsmanship of the builders shows in the skilled fitting of mortise and tenon notches of the frame. Two hewn

The Mabee Farm's barn is large by Dutch barn standards, measuring 52 feet long by 54 feet wide, with five bays. Walking inside the 250-year-old barn, you can see and feel Old World craftsmanship — huge, hand-hewn beams with wedged tongue extensions at most connections; wooden hinges on doors; and fat short braces, uncommon for North American barns. Although this barn was built sixteen years before the USA became a country, and much has changed since, most timber-frame building techniques are still based on the Dutch barn model — an amazing testimony.

posts, typically fourteen inches by fourteen inches, and longer than sixteen feet rise to meet the even larger horizontal anchor beam, from twenty to thirty feet in length, completing the unique *H* frame of the Dutch barn. The adzed anchor beams, typically more than twelve inches wide and twenty inches deep, protrude through the posts an average of eighteen inches with rounded tenons on the ends. Farmers used the center aisle for threshing the all-important grain. Opening both wagon doors created a draft to help separate the chaff from the grain when the sheaves of grain were flailed on the threshing floor. Spacious center aisles give plenty of room for wagons to enter, unload, then exit by the other door.

The side bays which faced into the central aisle provided convenient warm quarters for livestock. Spaces against the walls let cattle walk out through the side door and gave the farmer room to clean up.

Above the threshing floor, sticks laid between the bents on top of the anchor beams held the hay or sheaves of grain, one of the reasons for the high-pitched roof. People have found, among the sticks, pike poles originally used to raise these great structures.

These barns were usually built as soon as possible, even before a permanent house, because the farmer needed the barn to store the grain, livestock, and produce that would sustain the family. In the meantime the family often lived in a simple log house, waiting until the craftsman laid out each timber, notched each joint, and pegged each connection with hardwood pins. Finally, neighbors came to help with the great barn raising. Hundreds would lift up each bent, join them with connecting girts, forming, as the day progressed, the timber frame that would be then roofed and sided with clapboard in the coming weeks.

Some of these barns can still found in the Mohawk, Hudson and Schoharie river districts of New York State. After the end of the War of Independence they were introduced on the Canadian side of the St. Lawrence River by United Empire Loyalists who had fled the conflict. Unfortunately, the building of the St. Lawrence Seaway in the mid-twentieth century brought the destruction of up to 50 New World Dutch barns.

Remarkably elongated barn-door hinges from a New World Dutch Barn built around 1790.

This view of the Salt Springville New World Dutch Barn, built around 1790, shows the holes cut out at the top of the barn for martins, birds that were useful in reducing insects.

New York Dutch Barn — Salt Springville

Here is a barn steeped in history. More than 210 years old, it was saved from almost certain decay and demolition by community-backed action during the hippie days of the early 1970s. This beautiful example of a New World Dutch Barn has survived neglect down through the years as well as the torches of Loyalists during the Revolutionary War of Independence.

Prior to settlers coming during the 1700s, natives frequented the area because of its salt springs, which attracted game. Through the present-day village of Salt Springville lay a well-worn native trail from the Mohawk River to the Susquehanna River by way of Otsego Lake.

During the Revolutionary War the site of this barn was used as a resting and feeding place because of a creek nearby. A wagon march led by General Clinton on the way to the Susquehanna River consisted of 400 boats loaded on wagons and 3,000 patriots. Although most of the Dutch Barns built in this area before Independence were burnt or destroyed, this one, built around 1790, served as a working barn for 185 years until a depressed agricultural economy and subsequent neglect led to its deterioration.

In 1972, many young volunteers led by Louise P. Moore lovingly restored the barn to its original condition. By 1976, the National and State Register declared it of historic significance. Today the barn is used as a center for community meetings, private parties and a gathering place for the neighborhood.

An impressive example of a wedged tongue extender at the ends of the anchor beams.

The interior of the magnificent structure has four great anchor beams tying the bents together. Each bent, made from pine and oak, weighed more than 5,000 pounds. Oxen assisted the men in pulling up the bents at the barn raising. You can also see inscriptions and names on the anchor beam, made when the hay was piled on top of saplings and young hired men would scribe their names or doodle while the owner made the deal with the hay seller.

The timber frame and 35-foot rafters are all original. One can still see Roman numerals cut into the beams to identify each piece when the frame was put together so long ago. Signatures and names on the anchor beams date back more than 100 years to the period when the barn was used for loose hay storage. When dealers bought hay, records were kept with charcoal on pieces of shingle. Young hired hands, lounging during breaks on the anchor beams where the hay was stored, amused themselves by carving or writing their names and those of their girl friends on the beams with charcoal or knives.

As the economy changed, the barn was converted for a dairy operation. Two additions, one in front and one in the back of the barn required sawing off many of the great posts or columns. A new partition separated the bays. After dairying ceased, the barn served only to store hay and fell into disrepair.

Louise Moore consulted Professor John Fitchen, an expert on New World Dutch Barns, before beginning restoration of the barn in 1973. Volunteers first removed the huge quantity of loose hay that covered the whole floor and filled the south bay. In the next few years, they rebuilt crumbling foundations, spliced rotted sills, removed both additions and all siding and roofing. They straightened and repaired the heavy framing and made a new floor from two-inch pine planks secured with splines. The crew replaced some of the massive rotting beams and added a new cedar shake roof, all to original specifications.

The whole work of restoration was carried out by a team of carpenters all in their twenties, mostly hippies with wonderful insight into the past and a vision of the future.

New York Dutch Barn — Sharon

The man who built this barn survived scalping by a Mohawk during the Revolutionary War and being left for dead.

In 1776, Mohawks and Loyalists (Tories), who were leading in the fight against American Independence, took prisoner Jacob Diefendorf, a teenager at the time. Experience had taught the pioneers that the Mohawks would kill any prisoners they had captured if they were followed or attacked. The local militia followed and confronted the Loyalists and Mohawks near Sharon, New York, just west of Albany. The story goes that as Jacob tried to run away when the attack began, an Indian warrior jumped on his back and scalped him.

When the militia found young Jacob they thought he was dead but carried him back to the nearby town where he miraculously recovered, although his head was marked until he died in his seventies.

Just a few years after his horrible experience, Jacob built this New World Dutch Barn near the village of Currytown, about 10 miles east of Sharon. His effort was part of the rebuilding of the whole town and area, where many buildings had been burned during the time of Jacob's capture.

Two hundred years later, in 1975, when the barn was no longer being used and had fallen into a dilapidated state, John Wiggen bought the building, dismantled it piece by piece, and moved it to his property in Sharon — ironically the location where Jacob had his momentous encounter with fate.

Another interpretation of a pinned tongue extender.

New York Dutch barn, Sharon, New York.

Graffiti more than 130 years old testify that J. Burton was here in 1877.

John slowly pieced together the giant jigsaw of timbers and added a new roof and fresh siding. At this time the barn became the business home of the historic Johnson National Insignia Company, owned by John's brother Leonard, who moved there from New York City, where the company had been incorporated in 1895.

Within the barn the timber frame shows its fine original shape, including those Dutch barn signatures, anchor beams with huge *H* frames.

The barn is wider than it is long, as are most Dutch barns. It has three bents in total. The two main wagon doors, one in front and one in back, are gone, but the steep pitch of the roof remains true to the barn's origins.

One of the 12-inch by 18-inch anchor beams carries some fascinating graffiti dating from the 1800s. History seems so much closer when we see evidence of direct human touch such as hand writing in a place where ordinary people worked and played.

You can still see the Roman numerals used to identify each beam as it was put together more than 215 years ago.

A small, very red barn in North Harpersfield, New York, startling against the complimentary green of the ski slopes of Mt. Utsayantha.

English Barns — INTRODUCTION

The rectangular, vertical-boarded barn is still the most common agricultural building people see as they drive along highways and country roads in eastern North America. The English barn took this earliest form, becoming the standard for most barn frames in the eighteenth and nineteenth centuries.

Also known as the three-bay barn or threshing barn, this building was central to England's agricultural economy in the seventeenth century. Such barns had roofs of reed thatch and walls of stick and daub. Their most important function was to store the primary harvest of the farmer: wheat for bread.

When pioneers introduced these barns to the New England States and parts of Eastern Canada in the seventeenth and eighteenth centuries, wood shingles replaced the thatch and vertical and horizontal barn boards replaced stick and daub walls. Wheat quickly became a major crop, and so the English barn well suited the agricultural economy of these areas.

Modest-sized, compared to some of the later versions, early barns were typically 30 feet wide by 40 or 50 feet long, expanded in northern areas to 40 feet by 60 feet.

The three-bays-long, one-bay-wide structures, built on grade, had a central wooden threshing floor, hay storage in one bay and an enclosed stable for livestock in the other bay. The latter was a North American innovation; England's milder weather allowed keeping livestock in a separate smaller outbuilding. Above the threshing floor, boards or sticks were laid across the bents, as in the Dutch barns, so that more hay or stooks of grain could be stored there.

An English-style barn, typically built on grade, has three bays, the middle bay used for threshing grain.

ENGLISH BARNS

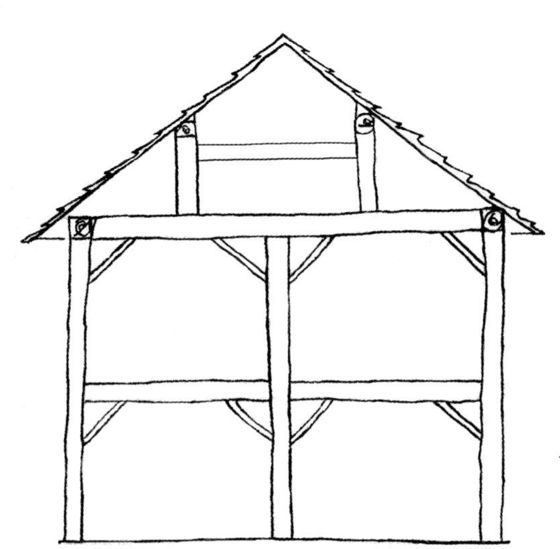

A typical side view of an English Barn frame, featuring three vertical timbers and horizontal girt making up one bent with a queen post roof support.

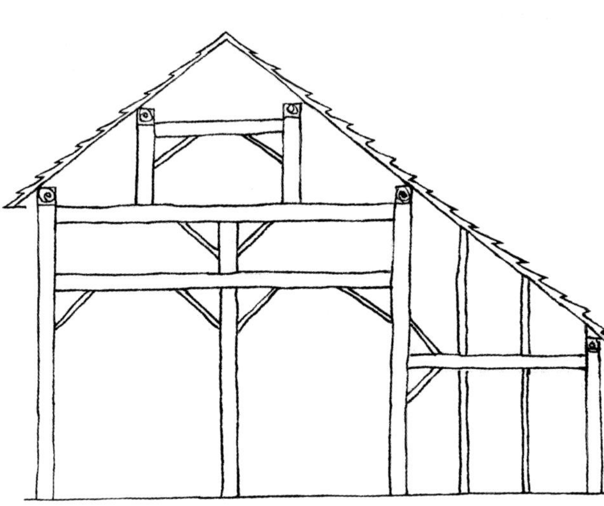

An English Barn-style bent with added braces and addition on one side giving it the distinctive salt box shape.

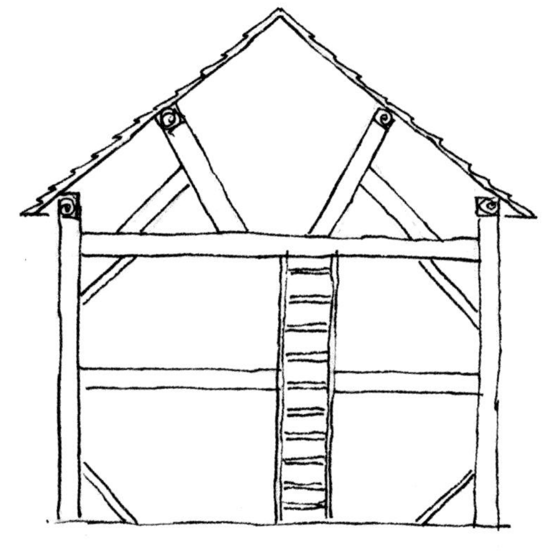

Another style of English Barn frame with a ladder built into a vertical post and slanted queen posts to support the roof.

Here's a practical, inexpensive idea: build a wooden silo and protect it from the elements with asphalt shingles. This neat, small barn near Potsdam, New York, at one time was the house and connected English barn.

The accent is on furniture and the colour red! Barns have been converted all around the country, some to restaurants, many as antiques sales barns, and this one, in Grahamville, Vermont, for furniture sales. The original barn with the cupola dates to the late nineteenth century. Aided by outside displays, the overall effect is eye grabbing.

Located next to a small creek near Tippecanoe, Ohio, this small barn doesn't seem well located for farming purposes today. The creek probably once watered livestock and, if the forests cleared a century ago left soils unsuitable for crops, the abandoned fields have grown up over the years.

The English barn was free standing, far enough away to pose no danger to the house in case of fire but close enough to be convenient for the farmer on his many walks to the barn every day. Yet this type of barn also played a part in another version of agricultural building construction — connected barns — found especially in Vermont and Maine. Early in the colonial era farmers used their English barns for threshing and the storage of feed only. When they needed more space, they constructed other buildings to house livestock, equipment, a workshop and firewood. These were sometimes connected to each other and to the house for easy entrance to all. Fire in one building was a constant fear, as all the others would surely have burnt as well.

A small, forgotten barn along the road in the Appalachia area of Ohio, near Freeport.

English barns were usually built with the sill plates resting on a low stone foundation or merely on a few rocks at the corners. As the agricultural economy turned more toward livestock by the 1850s, many barns were jacked up and a full, mostly stone foundation added underneath to house livestock.

The roof kept a high pitch, a carry-over from thatch used in England, where a steep pitch could carry away large amounts of rain. Here, the larger timbers inside were hand hewn from local trees, chiseled with mortise and tenon notches and pegged with hardwood pins. Ventilation came via cracks between the vertical barn boards, made in sawpits as

The Andersons' barn near Peabody, Ontario.

During the 1880s the Andersons' barn was raised in the traditional manner using neighbourhood help. The barn measures almost 35 feet to the peak on the ramp side and 45 feet on the river side; those men standing near the top of the roof were brave souls.
Courtesy of Andy Maier

This farm is set in bucolic surroundings but the river next to it made it a busy place for almost a century. The first of many sawmills was built here in 1872, harnessing the power of the North Saugeen River through a water turbine. Three sawmills burned down here either from lightning strikes or sparks from the metal sawing blade. The bustling enterprise had at one time nine houses to accommodate the hired workmen. [The last sawmill burned in the 1970s.] The typical English three-bay post and beam barn, built during the 1880s, is the only outbuilding left, although the dam and main house remain in place.

The inside of the Anderson barn roof shows its queen post supports and round rafters.

Where the barn boards and stone foundation meet, one can see the end of the large horizontal girt that holds up the sleepers and the main floor of the barn.

early as the 1650s. As more local water-powered sawmills began operation, horizontal siding replaced some vertical boards, especially in parts of Vermont and New York State.

On one of my trips to Vermont, I was waiting in a barnyard for the farmer to return, drawn down his lane by some very old-looking barns. He soon arrived in his dusty pick-up, having been to the local hardware store for supplies. We introduced ourselves and right away he wanted to show me his three connected barns, the earliest one built in 1790. "It's an English barn," said Willis Wood matter-of-factly.

His comment surprised me, for until then none of the farmers I had met had labeled their barns except by the date of construction or roof type. And there it was, the perfect example of an early English barn: thirty by fifty feet with three bays and some barn board still fastened with hand-wrought nails. The sills rested on dry stones about eight inches high. For a moment, I felt I had stepped back in time.

A well-kept three bay-barn with a large Victorian-style cupola, a common feature in the region around Wolcott, Vermont. The diamond-shaped window in this late nineteenth-century timber-frame structure is unusual but attractive.

The hills of the Central Leatherstocking region in New York State enhance this little white barn, across the road from a 1931 round brick barn, near Greene. The windows suggest there was a poultry operation at the front of the barn when it was built, probably in the late 1800s. Notice the large doors where there is no ramp, indicating an English barn that was later raised on top of a foundation.

Two crescent moons, a spade, and two diamonds adorn this barn that is more than 200 years old. The outside boards are furrowed from age, yet the English-style timber frame inside still holds up quite well. The barn and stables are stuffed with used lawnmowers and parts, part of the lawnmower and chain saw business of Jim Davis. His family settled this farm near Baxter, Pennsylvania, building the barn in 1805. Over the years, as the highway got busier, the Davis family opened a produce stand and then a small variety store in front of the barn. "They sold a little bit of everything," said Jim. He told me he misses those slower days, when traffic didn't whiz by and people stopped in more often, to buy something or just to chat.

I enjoyed the farms along the Connecticut River, the dividing line between Vermont and New Hampshire, one of the first places settled by pioneers during the late eighteenth century. I felt a sense of history in the area, with its beautiful brick homesteads built along the river and its fertile plains and hills pocketed with farms. This early timber-frame barn in Weathersfield Bow has some interesting transoms above the main doors and windows and has the longest-lasting roof material of all — slate.

A small English-style timber-frame building, built around 1825, originally used as a blacksmith shop in the little village of Edgeley, now forms part of the living museum at Black Creek Pioneer Village, near Toronto, Ontario.

A convenient arrangement of connected buildings just on the edge of Weare, New Hampshire. The house, carriage house, equipment shed, and barn are all joined so that the farmer could walk from his warm kitchen all the way into the barn without feeling the elements. Originally, the concept of the connected barn came from the province of Breton, France, and was adapted in this region. A connected barn has one big disadvantage: if one of the buildings catches fire, then all the structures could burn to the ground.

I've driven by this barn, built in the 1860s, many times and it always amazes me that the horizontal cedar siding has never been replaced. In places it's so thin you can almost see through it, but it's still defying the weather and storing agricultural machinery in its all-pine timber frame, near Hanover, Ontario. I was proud to have known the owner, Lloyd Chittick, a hard-working, generous man, always with a smile on his face. A "farmer's farmer" until he passed away in 2004, he spent his life steadily working the land, teaching me and many others the importance of being a good neighbour and serving your community.

Just east of Lake Champlain, the mountains of northern Vermont begin to take shape as wooded behemoths emerging in the skyline. Any valley or level ground is filled with farms settled in the eighteenth and nineteenth centuries by hardy pioneers. This English barn near Sheldon Junction may be small, but is a good example of timber frames in the region. The wooden silo is in great shape and not such a common sight any more since many have fallen to the enemies of rot and neglect.

Without a detailed local map I would have surely gotten lost on the narrow, windy, hilly back roads near East Craftsbury, Vermont. But it's these small dirt roads I love because on them I find unique, little-known barns tucked away on a hill, around a bend, or standing proudly in a cleared field.

This is Breezy Hill Farm, 2,000 feet above sea level, on top of one of the Green Mountains, a region dotted with ski hills and valley lakes. From here all roads and fields slope down and the wind must howl here during the autumn and winter.

I am convinced the barn was attached to the house precisely because of the location. Why walk outside during a storm when you could go directly from the kitchen into the barn? This direct attachment is unusual; other connected barns I saw generally put smaller driving sheds, now acting as garages or firewood storage, between the main barn and house. One more building separating barn and house would act as a barrier against animal odors and diseases, as well.

The many additions to this saltbox English barn show how the original timber-frame structure can support a great deal of weight and still stand for over a hundred years. All the additions, the woodshed, covered entrance way, driving shed and even the saltbox part of the barn are still being supported by the posts and beams which make up this original structure built in the mid-nineteenth century. It's located near Cookshire, Quebec.

Early English timber-framing techniques are evident in this barn, now a museum. Pioneer examples of gun-stock posts were made so that horizontal girts can rest on top of the post instead of being notched into the side of it with a saddle mortise.

This plain but pleasing barn was a typical English design when it was built in 1787. Luckily, the Windsor Historical Society came to its rescue in 1993, when the post-revolutionary structure was slated for demolition. The Society purchased and moved it from Weathersfield Bow to its present location north of North Springfield, Vermont, a distance of about 15 miles.

The blacksmith and horseshoer was a vital part of early rural settlements. This building, made from rough-cut two by four lumber, was originally a driving shed but has been converted into a blacksmith shop for the museum.

The pointy, wooden, slightly leaning silos bring memories of barns that have a fairytale quality to them. Now housing a pottery business near Taftsville, Vermont, these barns will probably never shelter livestock or store hay again.

This sturdy pine-framed barn is typical of Western Pennsylvania with its plain board siding and cut sandstone foundation.

A Poor Man's Barn

Allan Seigworth laughed as he told me that his sturdy, pine-planked barn located in western Pennsylvania was a "poor man's barn." Although not as showy as some of the expensive stone or brick ones found in the eastern part of the state it's still the most common type of barn in this region. And like other fertile lands in the eastern USA where hardy pioneers first cut the forests and built rugged timber-framed, no-frills "poor man's barns", this region has many that have lasted for more than a century.

Allan's three-bay English barn, built during the 1840s by the Craft family, was constructed from the towering pines that once grew at the back of his farm and throughout the surrounding Allegheny mountain region. After 160 years, the timbers still show red chalk lines used as base lines to measure from as the great logs were hewed with adz and ax into square beams.

These great pines were one of the first cash crops for farmers in this part of the country. After loggers felled the trees, farmers floated the logs through a series of rivers to Pittsburgh, the great city on the Allegheny River. Many of the men who worked the logs turned around at the end of the trip and walked back the 50 miles north to the Brookville area. They made the run again and again until summer came and the rivers became too shallow to float logs.

The Seigworth barn showing the beautiful queen post construction built more than 140 years ago.

This massive girt with a scarf joint is part of the timber-frame foundation holding up the 68-foot-long barn.

One famous pine log, Allan told me, was floated through the Susquehanna River system all the way to Chesapeake Bay and shipped to England to be used as a mast, a log 100 feet long and still a foot in diameter at the top. What a mighty tree!

Clearing the pines for fields led to the discovery of coal. Strip mining became and is still a common practice. Allan hates how such mining destroys fine farmland forever. He has heard stories of good barns being pushed over to get at the coal beneath them. "Only one generation made money," said Allan. One can still see piles of slag and ruined land around them. Then came oil discoveries, with farmers installing pumps in their fields, pumps left to rust when the oil ran out. The more recent discovery of natural gas has brought new pumps dotting the rural landscape.

Meanwhile, farmers like the Seigworths have tried to keep their farm in good heart, while making a living. They bought the 109-acre farm in 1962 and have developed a busy market for their produce. Their roadside stand sells sweet corn, potatoes, and other vegetables during the season. As well, Allan grows oats and other grain crops, which he sells to a local buyer.

Committed to agriculture as the backbone of a rural community, Allan feels cautiously optimistic, hoping to see rural areas once again depend less on big-city politics and become more locally managed, especially in food production.

Empty potato boxes sit waiting for the year's crop.

New York Three Barns

Only three barns remain on this farm from an incredible dozen originally built here, near Oxford, New York, beginning in the late 1700s when the first settlers arrived. The three timber-frame barns are well constructed and show signs of a once-prosperous farm and resourceful operation from the well-appointed buildings to the lavish use of cut stone for foundations. The longest barn, painted white, a variation of an English style, measures 35 feet by 125 feet. It originally stored hay and housed some livestock, including horses. A large cupola on the north part of the barn roof can be seen in the background of the archival photo taken at an auction around 1900.

The second barn, built in 1875, exemplifies wonderful craftsmanship. The high, split and cut sandstone foundation, includes such details as 4-pane built-in windows matching the ones above in the side walls of the building. The bank barn still looks in as good condition as in the archival photo.

The ramps to the double threshing floors are unique. Because the barn was not built into a hill, ramps had to be constructed to reach the main doors. These are carefully laid sandstone abutments with the last 20 feet spanned by two identical iron bridges, a costly endeavor for that time. The size of the barn, 40 feet by 100, makes it large by pioneer standards. The double threshing floors gave the farmer the freedom to take the horses and wagon in one door, empty his load, and go out the other without having to back the horses and wagon out.

The three barns in the foreground are all gone but the one in the back still stands with one cupola intact. The photograph, dated 1900, was taken from the iron bridge ramp of the blue barn. Some of the items up for bids are a fanning mill, a steel-wheeled work sleigh, a huge selection of horse harnesses, barrels, and a corn conveyer, probably operated by a belt from a stationary steam engine. Courtesy of Richard Place

One hundred and thirty years after it was built, this timber-frame barn still looks in good shape.

The last of the three barns, painted red, measures 30 feet by 70. Located behind the blue one, it originally stored agricultural implements and livestock. It has main doors on the gable ends, like Dutch-style barns, but has an English shape — narrow and long.

Although still a working dairy and beef farm, its owner, Richard Place, expressed disappointment with the current low milk prices driving many family-owned dairy farms out of the market. After more than twenty-five years in business, Richard told me that the price farmers now receive for milk is only slightly higher than when he started.

"People should consider saving local farms by paying reasonable prices for produce and milk and not base buying solely on low prices, getting milk from who knows where," said Richard.

This photograph of the finely crafted barn was also taken during the 1900 auction. Note the double threshing floors and two iron bridges leading into them. Two large cupolas used for ventilation top off the barn above each main door.

Courtesy of Richard Place

One of the two bridge-like iron ramps leading into the main barn doors.

Here is another view of the same auction with a nice steel-wheeled reaper binder in the foreground. Note the stepped down stone of the gray barn's ramp, from which some of the 1900 era photos, were taken. Courtesy of Richard Place

A wonderful collection of three timber-frame barns in Springfield, Vermont, each from a different era, beginning in 1790 with the middle English barn.

Vermont Cider Mill Farm

Tucked into the hilly backwoods of southern Vermont is Wood's Cider Mill Farm with three attached English-style barns. The oldest (middle) one dates back to 1790 — what a rich history on this farmstead.

The Springfield area was settled in 1759 but farmers found difficulty eking out a living on the stony, rugged land. Settlements and farms in Windsor County peaked between 1780 and 1820. When good land was discovered out west, many farmers and townspeople from around Springfield moved to those fertile plains with new dreams. Between 1820 and 1950 Windsor County slowly depopulated, its countryside and small towns dotted with abandoned buildings, including barns. By the middle of the twentieth century, urbanites from Boston to New York City began to discover Vermont for its breathtaking scenery, fishing, camping, and inexpensive houses and land. Today, tourism, skiing, cottages and retirement homes, rather than agricultural prospects have brought people into this state.

Originally, there were eighteenth-century log barns attached to the present three timber frames. Although the log structures are long gone, the timber-frame barns show an interesting assortment of pioneer building types.

Each gable barn is about 30 feet by 40 feet, one of the most common sizes for barns in England. Two have forebays as added protection for livestock. The two barns on either side of the oldest middle one were built in the early nineteenth century, using the same English barn methods of building. The middle barn still has hand-wrought square

Stones piled carefully on top of each other make up the foundation for this barn, now more than 200 years old.

nails — common until the introduction of machine-made nails about 1810 — holding some of the boards for the granary. Inside is a very unusual timber-frame innovation.

The late 1800s brought the invention of the hayfork, a time-saving device that hung from the top of the inside of the roof on a rail. A farmer could lower the fork, pick up the loose hay from his parked wagon then raise the fork full of hay and pull the load along the rail with a rope, to dump it at either end of the barn. The hayfork was revolutionary for its time, since hay and straw had been unloaded by hand from wagons since the Middle Ages.

In this barn the top horizontal girt stood too high to allow the hayfork and its load of hay to pass. So, sections of two girts

The bent girts were cut and lowered to accommodate the hayfork carrying loads of hay and straw.

The hayfork was a real time-saving device, as farmers didn't have to hand-pitch the forage into the barn. Courtesy of Archives of Ontario

were cut and lowered by about a foot to allow the hayfork to go over them. Cutting timber framing can weaken the overall structural integrity of the building but, in this case, although the hayfork has disappeared into history, the barn remains standing 100 years and more after the alteration.

I was also interested to learn that Wood's Cider Mill Farm has been making cider jelly and boiled cider for more than 120 years, still pressing juice out of apples with the original cider press. The twin-screw cider press was purchased in 1882 from the Empire State Press Company, in Fulton, New York. Water from a nearby stream ran it until the early 1900s. Today an electric motor and belt power it but a spoonful of Wood's Cider Mill Farm jelly is a taste memory of a hundred years ago.

This 1882 cider press is still in operation at Wood's Cider Mill farm.

Some of the earliest farms in Canada are found in the Eastern Townships of Quebec. In 1792, Lower Canada, as it was known then, was divided into Townships, each ten miles by ten miles. English settlers, who came mainly from the newly formed United States as United Empire Loyalists, could claim lots of 200 acres each. This rolling land, dotted with barns is near Waterville.

Quebec Long Barns

Introduction

THE PROVINCE OF QUEBEC is a unique cultural island in North America, where a vital people have kept their own language and a heritage that goes back to New France, founded in 1608 by Samuel de Champlain.

It is no coincidence, then, that there still exists in Quebec a style of barn different from any other in North America, influenced by the French countryside building styles where the settlers originated. When France lost on the Plains of Abraham, Quebec City, in 1759, British building styles also became a large influence, and many examples of English barns remain across the Quebec countryside. After the USA gained independence in 1776, some Quebeckers adopted American styles of barn building: Dutch, round, and later Pennsylvania German bank barns.

Quebec-style barns are marked by elevated main wagon doors, built into the low eaves of the roof with a large dormer. A smaller door for livestock is on the right and, between them the farmer's door, set off by square windows, all proudly painted a vibrant red. An old-style hay rake sits in front of the barn, located near Sainte-Famille, Quebec.

Along the shores of the St. Lawrence River, especially east of Quebec City, are some of the best examples of French-influenced Quebec Long Barns. These structures typically have low eaves, are narrow but long, with a main dormer door entrance to the loft, and many different sizes of doors and windows. Traditionally, they are painted a bright white with rich red trim.

Driving around the Île d'Orléans, just outside of Quebec City, gives a wonderful glimpse into the past. Settled originally by 300 French families in the 1700s, it's a microcosm of traditional Quebec looked upon as the birthplace of

Francophones in America. Only one main road winds some forty-two miles around the island. Small villages and farms crowd the shoreline road and the "garden basket of Quebec City," producing vegetables, fruit, berries, cereals and maple products.

Red and white barns dot the shoreline in many places, some dating back as far as the 1780s. Although there are small variations, the overall structure remains quite consistent. When he was governor of New France in the early seventeenth century, Champlain declared that barns should be narrow (20 feet wide), and long (40 feet or more) like those found in Normandy, France. He set the standard for the Long Barn that has been followed for more than 300 years.

Their most striking feature is the main door, typically ten feet by ten feet, opening to the loft. Loft access necessitated a dormer because of the low eaves, and usually a ramp. Built on grade, these barns house cattle below and hay storage above. Where barns don't have ramps, hay or stooks were brought by wagon up to the brightly painted doors and then pitched inside.

Another fine example of the elevated wagon doors of a Quebec-style barn, near Sainte-Famille.

The early barns were sometimes termed "connected," because of the many different functions which occurred inside, all under one roof. This at one time included living quarters for the farm family, but the practice stopped by the early seventeenth century when settlers built separate houses, many of them stone. Typically, connected barns have two to four mows on grade with wooden floors, one or two entrances for the storage of implements, and a threshing floor. The livestock, initially cows, pigs and horses, were placed in the middle of the barn, below the loft from where the hay could be easily dropped down. Windows set into the wall provided ventilation for the livestock's health and light for the farmer's work.

This Quebec-style barn has all the customary indicators — the dormered wagon doors; additional livestock door on the extreme right; a livestock and equipment door on the extreme left, high enough to let in horses, the farmer's door in the middle and windows where they were needed. Located near St. Pierre, Île d'Orléans.

A French farmer built this barn in the 1830s on Île d'Orléans. "He must have been poor by the way it was built," said Yves Robitaille, who now owns the farm. He pointed out the rudimentary poles used to construct the walls, spaced unevenly, and the minimal rafters used for the roof. Inside, small posts, propped here and there, once supported hay and sheaves of grain on the floor above. For now, Yves is not using the barn for farming. "You never know about the future, though," he said with a smile.

The loft above, filled with hay, kept the livestock warm. Barns were made from wood, with low stone foundations.

Barn builders used basic timber-framing techniques, but just as common were posts set in the ground with sawn lumber or more poles for walls. The loft area, extending less than the length of the barn, was supported either by a post and beam frame, or just with posts in the ground and cross posts for support. Thatched roofs were common until the middle of the nineteenth century, but straw and hay, the only available materials, were never as good as the reed found in France and needed to be replaced more often. After thatch, wood shingles became popular for roofing as well as for walls. The Quebec Long Barn was most often sided with tight vertical boards, relying on cupolas for venting.

Farmers in Quebec grew an ever-changing variety of crops, often affected by the political climate and treaties signed. Early colonists grew and stored corn, oats, rye, barley, buckwheat, hemp, flax, and feed grains. After the conquest of the French by the British, cereal production became primary as an export product to other parts of the British Empire. In 1830, after the collapse of the corn and grain market, peas and potatoes became important. In 1856 the US market opened up after the signing of the Elgin-Macy Treaty, and the keen demand for more farm products encouraged settlers to expand farther into Quebec.

Agriculture is still very much alive in Quebec and is certainly so around the village of Saint-François on Île d'Orléans. I stayed on this island for five days, waiting for the rain to stop so I could take some photos — not a bad thing to happen. I took my time driving around, always with the St. Lawrence River as the backdrop, seeking barns that I could photograph later and talking to people in my broken French about their lives and barns.

One was Yves Robitaille, a gentle, soft-spoken, wood craftsman, with a shop in part of his brown-shingled barn. He first wanted to show me inside his renovated 200-year-old stone house.

There is exciting history on this island in the St. Lawrence, just east of Quebec City. It was on the Plains of Abraham that French general Louis Joseph de Montcalm lost his epic battle in 1759 to the English general, James Wolfe, and sealed the fate of New France. Before and after the battle, English soldiers raised havoc in the countryside, including this farm's location. Their destruction did not spare the original barn. The stone barn foundation, on the opposite page beside the present-day barn, is all that's left of the structure burnt to the ground in 1759.

The hardwood timber joists sloped at different angles as the house had settled from age, and the wooden frames around the windows did not match the new angles of the walls. But he wanted to keep the old handiwork, because it was "made from the hands of men from a different time, and that is important to remember, to remember our history," he said.

The Robitaille barn, not particularly striking inside, was built in the early 1800s, from poles sunk into the ground with odd lumber used to support the walls. However, I saw an old stone foundation beside the barn, showing where the original barn used to stand. When the invading British army of James Wolfe came in 1759, this barn was burned to the ground by the British troops, as were many, many others. That explains why few of the original barns dated before the invasion are left. Almost 250 years later, the stones of the foundation remain, placed by the hands of an early settler consumed with a dream to farm in the new land.

The use of colours, characteristic in many older barns in Quebec, makes this plain barn near Saint-François stand out. It has the ramp with dormer on the other side and the small door for livestock and added tracked door for machinery on this side.

French settlers from Normandy brought with them a characteristic framing method known as the king post. This type of post and beam structure used a typical bent. On top in the middle, it added an upright timber called the king post. A horizontal beam, the collar tie, ties the king post to the principal rafters on either side. Even after New France was conquered in 1759, both French and English continued to use the king-post timber-frame method well into the 1800s, when the queen-post timber frame became common. Queen post refers to a pair of vertical timbers coming from the top of the bent, which join to the purlins to support the roof.

This incredible red and white barn was built in the 1790s using the king-post method. On this dairy farm, cows are milked in the stables as they have been for the last couple of centuries. The loft area is indicated by the windows, where the stables are located, with the full height of the barn on either end left for equipment storage.

Deceptively unobtrusive, this small white timber-frame structure was built almost 250 years ago and the stone house shortly after. Originally a coach house, it was also used to store hay and agricultural equipment during its long history on Île d'Orléans in Quebec. It has the unique king post framing style brought over by settlers from Normandy, France.

Some of the oldest barns west of Carlisle, Pennsylvania, were built along Creek Road, a quiet, historic road which brought settlers to their bit of paradise. This very narrow-slitted stone barn probably dates between 1790 and 1810. At that time, farmers were beginning to make a reasonable amount of income and were able to afford to get rid of their small pioneer-built log barns and construct something that would last centuries. The ceramic silo is common to this area and much of Pennsylvania as well as Ohio, New York, and Michigan.

Bank Barns

Introduction

ONE OF THE MOST UBIQUITOUS farm buildings dotting our landscape is the bank barn. Big and practical, its origins are found in the German and Mennonite Swiss farmers who brought to North America building concepts from the mountainous regions of Bavaria and the Swiss Alps. These two-level barns, also known as the Pennsylvania German Barn, housed in one building everything the farmer would need.

In Old World Europe, farmers typically had their stables, haymow, and living quarters under one roof. Built into a hill, the barn had an entrance at ground level into the stables and also one from the hillside into the haymow. Taking this idea to Pennsylvania, settlers during the mid-nineteenth century continued to build into the sides of hills. One long wall preferably faced south, leaving the north side set into the hillside. This arrangement provided the farmer with a full basement or stables for livestock and a second floor to store hay and grain. He could enter the stables from ground level and the haymow through large wagon doors. If the hill wasn't quite high enough a ramp would be built to gain entry into the haymow. The idea of living within the barn never took root in North America, except for a few early instances.

Unlike English and Dutch barns, the bank barn divided the crops from the livestock, still giving the farmer enough space to thresh grain, store crops and stable livestock all under one roof.

Another characteristic feature of this barn was the forebay, an extension of the barn floor (like a bal-

The classic Michigan-style decorated main doors found on many barns throughout the state. Pioneering farmers were often both practical and religious and any kind of fanciness was not looked upon kindly by the strict Christian codes of those days. These white lines around doors and openings were meant to "keep the devil out."

cony), the length of the barn, past the stone foundation and enclosed as part of the haymow. The forebay not only added to the capacity of the hay mow but the extension also gave protection for livestock that could stand underneath the overhang during inclement weather or for shade during hot sunny days. At first forebays were cantilevered structures using huge timber joists, called sleepers, spaced closely together to support the weight of hay on top. Later, near the end of the nineteenth century, as large trees became scarcer, wooden posts or concrete pillars added under the outside corners held up these extensions.

Some historians reason the forebay was imported from parts of Europe where houses built during the Middle Ages had short second-story extensions. As well, some Swiss barns had large overhangs for storing wood and keeping the walls dry. Not all bank barns were built with forebays; some had just a small shed-roofed overhang, perhaps only two feet wide, called a pentice, to protect the walls below.

A large timber-frame barn with a typical farmer's no-frill cupola for air ventilation on the roof and open-air stalls in the stables located near Brisben, New York. The open-air stalls were once part of a forebay, so common in early bank barns.

Inside bank barns, the timber frames rose high, often reaching forty feet to the roof. Such beautiful unadorned vernacular architecture is living proof of the amazing craftsmanship of this era. Beams often twelve inches by twelve inches, mortise and tenoned and pinned with hardwood pegs were held together at every connection with forty-five degree braces. Like other pioneer barns, bank barns were built from whatever trees grew on the farm, or close by, so that frames made from only one kind of wood are unusual. Builders valued rock elm or oak for sill plates, maple for posts and pine for girts, but also used beech, ash, butternut, hickory and even hackmatack or larch (tamarack). Typically, they cut trees during the winter months. With adzes and framing axes, they squared logs in the bush while still green and easy to chip.

Overgrown with pigweed and cluttered with rusting implements this timber frame barn stands near Greene, New York. The long narrow profile might indicate that it was used for pork production, as small animals require less room. During the 1800s, everyone had a mixed farm operation which would have included a few dairy and beef cows, poultry, and a large garden.

A Victorian bank barn beautifully nestled into the hillside in the grape-growing region of Lake Erie. This is a long-settled area, juxtaposing historic barns like this one next to clusters of new and century-old homes, just outside North East, Pennsylvania.

Once the hillside was sliced out, stones that had been gathered were carefully laid in place for a solid foundation. The form depended on the stone available on the farm and the type of mason. Masons usually mortared fieldstones into two-foot-thick walls from six to eight feet high. Scottish masons, especially, constructed beautiful coursed or split stonework, an attractive and long-lasting foundation.

With the stonework and the first floor deck finished, the squared timbers were brought onto the first floor and the notching began. A master framer outlined all the joinery that needed to be bored and chiseled. Apprentice framers or experienced hired help, often young farmers starting out, finished the notches. As the original great forests were cut, craftsman framers began to use different notches, such as scarf joints, to splice the beams together. These were useful for purlins and top plates, where timbers needed to span the length of the barn, many more than 60 feet.

After finishing the joinery, the framers assembled the great bents and pinned them with oak pegs on the deck in the order that they would go up. Then neighbors and the community came together to erect the barn structure, an event often called a raising bee.

In parts of Pennsylvania, gable walls, or sometimes the ramp-side barn walls, were built from stone or brick. Often the farmer could find some clay on his land to make bricks. Since ventilation was very important, on stone end-walls embrasures were built in. Beautiful brickwork lattice patterns served this use on brick walls. Forebays were always built from wood as bricks and stones were too heavy for the cantilevered structure.

The beautifully symmetrical cupola on the red barn has west and east marked on the iron wind vane and a glass globe on the lightning rod.

Initially, barn roofs had rafters still round from the tree, boarded on top and then covered with either slate or wood shingles.

Bank barns were designed to have ample threshing space, many with two or three threshing floors and a corresponding number of wagon doors. As technology improved and mechanized threshers and later combines became popular, the threshing floor became the drive floor, used to store farm equipment, including combines and tractors. Imagine the tons of hay and grain stored on the first floor, and then equipment as well, all held up by the posts underneath, the large sleepers and the stone walls of foundations built more than a century ago.

By 1900, bank barns were the most common design used in this part of the country. Many converted English barns by jacking them up, putting a foundation under, and adding a ramp to the first floor. When the West was opened up in the late nineteenth century and took over wheat production, bank barns, which had been built with an emphasis on grain and hay storage, were adapted to greater livestock production, which became increasingly more important in the East.

The fresh white paint on this big, well-kept T barn with matching white wooden fences shows that this is still a working farm, near Kidron, Ohio. Additions to barns were typically built in a T or L shape so that the two buildings could be connected with a common door.

Demand for farm products, including meat, continued to rise, especially during the period of the American Civil War from 1861 to 1865. Barns grew larger. Farmers added other smaller buildings, such as a workshop to fix the many implements, a woodshed, a wagon house (later called a drive shed when tractors were introduced), and a corncrib. As the economy further changed by the 1950s milk houses were added for the important dairy industry as well as poultry sheds for the new mainstays of the agricultural economy in Eastern USA and Canada.

What started out as the Pennsylvania German Barn in 1800 became just the Bank Barn by 1900, as ethnicity and its associated building styles became less and less important and an indigenous architecture evolved from a combination of different cultural influences and economic and technological innovations. Still the most commonly seen barn, they are classics of eastern North America's landscape.

Living in the rural area of Southern Ontario, I pass by bank barns every day, many still used for dairy, beef, and sheep production. Simple and unpretentious on the outside, their real beauty is inside, in the great timber frame structures that soar to the roofs and lock together with simple but enduring post and beam connections.

In late summer, the golden stubble of oats and barley testify to a good harvest. Owned by my neighbours near Chesley, Ontario, this 110-foot by 40-foot timber-frame barn was built into a hill in the 1860s and hasn't seen much renovation since. Two ramps lead to a double threshing floor. Inside the barn, the sun's rays passing between the barn boards play tricks with light and shade on a queen-post timber frame which is still in great shape.

Along Highway 28, southwest of Brookville, Pennsylvania, one of the 10,000 Mail Pouch Tobacco barns painted with this advertisement between 1920 and 1969.

Mail Pouch Tobacco Painted Barns

This sign is a familiar and frequent sight across the Midwest and parts of New England, thanks to the advertising genius of The Bloch Bros. Tobacco Company, incorporated in 1890. They began applying their famous signs to the sides of barns in the 1920s, painting 10,000 of them by the time the business ended in 1969.

Typically, the company would send a crew — the painter and an assistant — out on the road for months at a time. The two men would have some pre-assigned barns to paint, but often they just knocked on doors and asked farmers if they wanted the sign on their barns. Each of the black, yellow and white signs took up to a day to complete, depending on how old and dry the barn boards were and how much paint they soaked up. Apparently, the painters always started in the top center of the sign, with the letter *E* in CHEW.

After the Second World War, a painter's salary was about $32 a week, out of which he needed to pay his lodgings. To save money, the crew often slept in their truck or in a farmer's barn.

The farmer had a choice of payments. He could get a lump sum of cash, magazine subscriptions, or a hefty supply of Mail Pouch Chewing Tobacco. Sometimes he could pay for the rest of the barn to be painted, but it had to be done with the colors on hand: Mail Pouch Tobacco black, yellow, or white!

Another two miles down Highway 28, this prominent barn, though sadly neglected, still sports a re-painted logo of the famous tobacco advertisements for Mail Pouch Tobacco.

CHEW
MAIL POUCH
TOBACCO
TREAT YOURSELF
TO THE BEST

The last barn of the 3,000-acre Boldt farm and a many-windowed addition that resembles a small house but was probably used to house poultry. The main barn has unusual diamond-shaped cupolas and a gabled entranceway into the threshing room.

New York Boldt Farm

Here was a builder who thought "Big" and "Grand." Having built the Waldorf-Astoria and Bellevue-Stratford hotels in New York City, multi-millionaire George Boldt gained the status of organizing genius. His farm on Wellesley Island, New York, gave further proof.

The farm, built in view of his incredible castle and yacht house on Heart Island in the 1000 Islands region of the St. Lawrence River, was impressive by any standards. Encompassing 3,000 acres, it showcased a modern enterprise that most farmers of the late 1800s could only dream about. At one time it ran the largest poultry operation in New York State, raising thousands of geese, turkeys, chickens, ducks, pheasant and quail. In 1900, using a refrigerated railway car from nearby Alexandria Bay, Boldt shipped to the Waldorf-Astoria Hotel 1,864 pounds of dressed Wellesley turkey all prepared at the modern butcher shop on the farm for Thanksgiving. To get to the railroad station on the mainland, he built an electric trolley system from Wellesley Island. The farm also shipped butchered lamb, pork and beef as well as such dairy products as milk, cream, and butter to New York City.

He employed dozens of people on his farm, which also housed hundreds of polo horses for his rich friends' entertainment. He also developed a golf course.

Unfortunately, when his wife died in 1904, he abandoned the famous castle and sold the farm.

The Boldt farm as it looked in 1900 on Wellesley Island in the St. Lawrence River. Today, all that is left of this complex is the elegant barn with its steep cupola and the stately house. The farm supplied meat and dairy products, shipped by refrigerated railway car, to the Waldorf-Astoria Hotel in New York City. Courtesy of John Holcombe

George Boldt built the 120-room castle for his wife Louise, who died suddenly in 1904. All construction stopped and Boldt never returned, leaving the castle to the elements and vandals, and selling the farm. In 1977 the Thousand Islands Bridge Authority acquired the castle and has spent millions restoring the famous property as a landmark for future generations to enjoy.

Ohio Amish Country

The Heart of Amish Country, Wayne County, Ohio

Wayne County, Ohio, and the surrounding counties are home to one of the largest communities of Amish in the world. The efforts of Amish people to keep a simple way of life and maintain self-reliance stem from religious principles established in the 1500s as part of the Mennonite religion. Their mostly German-Swiss descendants have foregone the technologies and conveniences of modern life.

Amish still travel in their distinctive black horse-drawn buggies and rely on wagons for hauling material and livestock. Farm animals and the barns to house them are central to their rural life. Most of their barns are bank barns, especially the new ones. Although they try to be as self-sufficient as possible, the Amish also help each other as a community through barn raisings for young couples starting out or after disastrous fires or for the construction of a new school or church. They attend one another's quilting bees, threshing bees or just plain help each other as neighbors. Farming has always been the ideal Amish chosen work, as they feel the land brings them closer to God, who created it. Farming keeps the father at home and the rest of the family joins in the endeavor, bonding all members in a common undertaking.

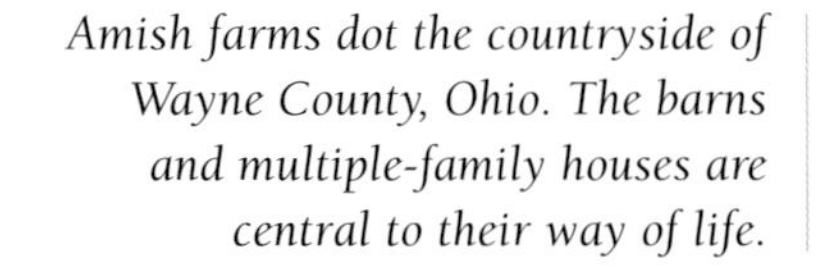

Amish farms dot the countryside of Wayne County, Ohio. The barns and multiple-family houses are central to their way of life.

There are always Amish buggies lining downtown Kidron, Ohio, a town with a population of 150. This center of Amish commercial community provides restaurants, an auction barn, and Lehman's Hardware Store.

The village of Kidron is the heart of the Amish community in Wayne County, where the auction barn attracts thousands of people every summer to view and buy antique but still usable horse-drawn equipment. Watching Amish farming techniques is like traveling back to the early 1900s. They still use reaper binders, threshers, and one-furrow plows, all powered by sturdy draft horses.

Kidron is also home to Amish restaurants serving simple home-cooked meals and, of course, Lehman's Hardware store. Jay Lehman, who founded the store in 1955, wanted to provide the local Amish and other self-sufficient folks with products that didn't need electricity. Today the store is the world's largest supplier of nineteenth-century technology: oil lamps, wood-fired stoves, butter churns, apple peelers, gas-powered refrigerators and just about any old-fashioned tool you can think of. The Amish, who were core customers for so many years, have now also become suppliers to the store, bringing handmade quilt frames, well buckets and wheel barrows.

I see the Amish driving around in their horse-drawn buggies every day where I live near Chesley, Ontario. I always wave to them and they wave back. Neighbors with biblical first names such as Noah and Gideon repair my leather working boots or saw lumber for me. The children are a wonder, full of play and always with a friendly wave. They walk in groups to school, the girls in bonnets and dark blue full-length dresses, the boys in black felt or straw hats. It's reassuring to see the Amish live their life without the need to copy the hustle and bustle of our complicated technological lives. Hard work based on sound farming principles and the joys of raising many children, who grow up in turn to help the family — seems nice and simple.

Ontario Leith Barn

Twelve 40-foot posts, all hand-hewn from maple, rise majestically to meet the purlin roof supports on this barn located near Georgian Bay, Ontario. This farm reflects a typical progression: the first log barn was built in 1847, then a small one-story English-style timber frame was erected in 1874 and finally, ten years later, the 100-foot long permanent bank barn we see today, built to last centuries.

"We did nothing else but farm but we all managed somehow," said Jean Bye, who with her late husband, Lester, began to farm here in 1944, using only horsepower. "Farming was so different then," Jean explained. "We always had the neighbors to help us and we helped them."

The community and neighbours all came to help with barn raisings. The Leith barn, a very sturdy structure, was raised in 1884.
Courtesy of Jean Bye

More than 130 years later, the Leith, Ontario, barn is standing the test of time and still an essential part of a beef operation.

Each bent of the Leith barn has two 40-foot posts that meet the purlins.

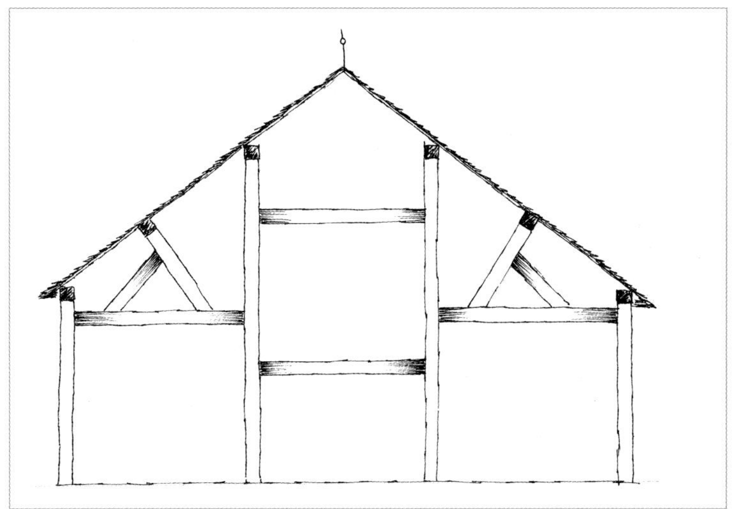

A full-time occupation, farming provided Jean and Lester with their own food and a decent living. They raised different types of livestock. All those animals needed lots of feed and, although it's hard to imagine, the huge barn used to be stuffed to the rafters with loose hay and straw every winter. "It was a lot of hard work," remembered Jean. A good source of cash, she told me, was selling cream to a local dairy. Every day's cream was stored in an old well in the basement to keep cool until they delivered the full can by horse and wagon to the dairy 10 miles away.

Ken Mitchell, who took over the farm in 1986, is proud of the barn. "There's good workmanship here," he said, pointing up toward the graceful beams and braces crisscrossing each other like a farmer's opera house, with Ken as the conductor.

Ontario Turner Barn

In the once-prosperous little village of Scone, tucked away in the hills of Grey County, Ontario, stands a large barn with a plain exterior. But inside, it features an incredible twenty-foot-high bridge running its whole length. This structure, built in the 1860s, was once part of a thriving town that included five water-powered mills, a hotel, general supply stores, blacksmith shops, a lumber and brick yard, and school. Today, all that's left is one mill, which produces electricity, a few houses and, up on a hill overlooking the rest of the village, Grant Turner's barn.

Grant, a gregarious, weathered local character told me the story of his barn, with many anecdotes about the prices of cattle over the last 20 years, and how farming has changed during his lifetime. His beef enterprise remains, as always, on a shoestring budget with "making do," and eating his own butchered meat, farm-grown potatoes and other produce. With his beat-up GMC diesel pick-up truck, he ferries animals, logs, scrap and firewood around, taking on odd jobs to supplement his income.

From the outside it looks like a typical Ontario bank barn, although it is very big: 114 feet long by 90 feet wide. But where there would usually be large threshing doors to enter the barn, this barn has instead a ramp at the gable end leading up, up, to the brainchild of the builder: a third-floor, 12-foot-wide bridge that spans the full 114-foot timber-frame building. In the days of loose hay, huge amounts were brought into the barn to be pitched up into the mow. This 20-foot-high bridge enabled hay to be brought by wagon and

This is the lower of two ramps leading to the third-story bridge inside the 114-foot barn.

horses up into the top of the barn and then pitched down into the mow. Similar to the layout of some three-story round barns, the bridge must have saved a lot of work.

To support this heavy bridgework and the large wagonloads of hay pulled in, initially by oxen, then by horses and to this day by tractor, beautiful sets of double-posted bents were built. The bents were assembled in the shape of a big *H*, with the bridge laid down on the horizontal part of the *H*.

Using the high bridge and the two ramps leading up to it on either side of the gables was not without incident.

Grant told me the former owner, Fred Lowe, had pulled a large load of hay with two horses onto the bridge inside the barn. It's generally understood that horses don't like being on bridges and Fred had trained his animals using blinders. On that day in the early 1900s, one of the horses slipped and fell off the side of the bridge. It was attached by its harness to the other horse and to the wagon and luckily this broke its fall but the poor frightened animal hung in the air, suspended over the haymow, kicking and whinnying. Fred whipped out his trusty pocketknife and cut the harness. The horse dropped like a stone but luckily landed on the soft loose hay below. Fred ran down to the second floor and, seeing that the horse was on its feet again, led it out of the barn. The horse was certainly scared, but had no broken bones, and so got the rest of the day off in the pasture, probably nursing a few bruises!

The other bridge incident, which occurred in the late 1950s, had a happy ending as well, but could have been disastrous. On the bridge, Grant's father and brother had just unhitched a big wagonload of loose straw, which they had brought up with a 1949 Ford tractor. They

The third-story, 20-foot high, 114-foot long bridge inside the Scone barn.

The incredible 20-foot-high ramp on the east side which the owner still drives on with his pick-up truck or tractor.

were both on the tractor driving out when, at the very end of the bridge, just before going out of the barn onto the outside ramp, the unthinkable happened. The main sills that crossed each 12-foot span of the bridge slipped from the edge support. The 3,000-pound tractor dropped, rear-end first along with that part of the bridge and floor onto a door jamb sticking out below. That broke its fall somewhat. Then the tractor pitched onto a huge mound of loose hay stored there for the winter. The father hung on for dear life to the steering wheel while Grant's brother held to the loader. Both escaped serious injury, although they were covered in decades-old dirt and dust accumulated in the old boards that had fallen all around them. The rest of the family was down below, at the other end of the barn, feeding grain to a noisy threshing machine, and had not heard anything. The hapless tractor passengers crawled down from the haymow and emerged onto the threshing floor with only their eyes and white teeth gleaming through the old dirt.

The tractor had to be let down the rest of the way with pulleys and all the wheels taken off so that the machine could be pulled out of the barn through a narrow opening onto a temporary ramp.

Grant showed me the spot on the bridge and how he had fixed it — large metal cables wrapped around the sills, attached to other larger beams with bolts holding them in place, just in case.

Since Grant bought the farm in 1952 he has built new ramps leading up to the bridge inside the barn, ramps with gentler slopes. This way he is able to drive his pick-up truck inside when he has a load of hay to drop over the side. He admits both ends of the ramp "need working on," since there are boards that are rotten and gaps left by others that have rotted through altogether. And, as with many old barns, this 150-year-old building needs jacking up where the stone foundation has crumbled and roof repairs where the snow load has broken some rafters. But it's amazing how these old structures can withstand years of foul weather with minimum maintenance and still hold crops for generations of farmers to come.

Vermont Bread & Puppet Barn

You would never guess that the three floors of this weathered 1863 Vermont barn are full to the brim with large and small puppets because the outside is like any other barn in the Green Mountains around Glover. Garbage men, crones and politicians, salesmen and butchers, soldiers and saints are all part of one of the world's biggest collections of puppets, exhibited in the Bread & Puppet Museum barn. The barn is home to Peter and Elke Schumann's Bread and Puppet Theater, formed in 1959 and moved to this farm in 1974.

The groups of puppets inside the barn are figures and faces mounted from many past productions and are in effect a history of the anti-war, anti-poverty and racial equality movements of the USA.

The Bread & Puppet Museum's brochure calls Peter's style a "prodigious mix of Romanesque, German Expressionism, Cycladic Minimalism and Potato-Nose Expressionism." His inspiration is "an expression not only of the accumulations of time but of the urgencies which inspired the making of so much stuff: the poverty of the poor, the arrogance of the war mongers, the despair of the victims, and, maybe even stronger than that, the glory of this whole God-given world."

Because of the ample space inside its tall timber frame, a barn was a natural choice to house all these past and present achievements. Also, since the Schumanns live in the lovely old house on this farm, the barn and its wild collection, like an exhibition at a regular museum, is a permanent place open to the public. The puppets and the barn suit each other in a peculiar way, both exhibiting creativity and antiquity.

The inside of the historic timber frame barn and its amazing puppet display.

The stalls and stables where cows were once milked and horses fed now house puppet displays. In the massive hayloft are the biggest puppets, some as tall as 18 feet. The barn's ceilings and interior walls are hung with faces and figurines. The collection is a stimulus, not only for artists, but as an alternative use of the vast space offered by the historic building.

I have visited this museum and the puppet theater many times since I first drove there in 1980. The magic held in them is hard to define, but I always come away inspired, challenged, and feeling that one person can make a difference.

The Bread & Puppet Museum barn, near Glover, Vermont.

Once livestock housing, these stalls now shelter magical inspirational works of puppetry.

I've taken this route dozens of times since the 1970s and I love the curvy roads, picturesque heritage towns, and historic farms sprinkled across the mountainous landscape. This farm near East Berkshire has everything that makes the best part of Vermont — a good crop of corn, a forest, and mountains in the background. The barn is in good shape with a nice cupola and new barn board.

Scottish stone masons built this massive stone barn in 1848, near Rockwood, Ontario. Many Scottish families immigrated in the 1840s to Upper Canada (Ontario) and built numerous stone buildings in nearby Fergus and Guelph. The stone walls of this English barn measure 34 feet wide by 60 feet long and 18 feet to the eaves. Still plumb as the day it was built, each corner was laid with quarried stone from Fergus. The walls are connected with girts spanning the width at the top and there is a central pine girt inside that is almost thirty inches in diameter, pine cut from virgin forests. Courtesy of Pief Weyman

The lower T of this barn, near Markdale, Ontario, was meant to store straw, built narrow and long with a hayfork track down the middle to throw the loose straw off to either side. The main barn still has the litter carrier bar sticking out of the doorway. Introduced in the 1920s, the carrier was used to take manure out of the barn without having to fork the stuff into a wheelbarrow and then push the heavy mass outside. The equipment used a single track bolted to the ceiling of the stables. The track followed the aisles with a wooden or metal half barrel in which the manure was loaded. Once full and hoisted up by means of chains, it was pushed along the track, outside the barn, and dumped onto a growing manure pile. A real time saver in its day!

Symbols were originally painted on the side of barns for good luck or to ward off bad spirits, as was this sun wheel hex on a barn near Markdale, Ontario. It was said that hex signs on the barn would drive away any evil spirit roaming around the farmhouse.

The timbers of this three-bay, 50-foot by 60-foot barn were all made from rock elm, an incredibly hard and heavy wood. The local paper from 1905 states that the barn frame was raised with the help of more than 200 people. Notched by a master framer, "there was not a mistake of misfit in the big structure."

Unfortunately, on the day of the raising, a main plate fell, hurting three men, one seriously. Those kinds of accidents happened frequently at raisings because, after all, it was hazardous work.

The Wellington County Museum barn.

This well-kept bank barn was originally built in the 1870s for $1,483 as part of the Wellington County House of Industry and Refuge near Elora, Ontario. This institution provided shelter for the "deserving poor," the aged, and the homeless until 1954, when it became a museum. The barn was meant to provide work and food for the people living in the House.

It's interesting to note that because the barn was part of a government institution, records were kept of the barn repairs or alterations. These included the cement silo added in 1914 and the enlargement of the stable area in 1937 with new steel stabling equipment installed at the same time. In 1942, the roof was re-shingled with cedar shakes while in 1952 new eavestroughing was added. Repairs were made to the foundation in 1979 and the barn was painted. The silo got a new roof in 1994 and the barn was painted again in 1996. It's wonderful to see the preservation of historic barns such as this one, and the equipment inside.

The cement silo was built in 1914.

Corn stooks drying in the late autumn sun. Amish harvest grain corn by hand, picking the cobs and throwing them on wagons. The corn is then taken to the barn to be stored for winter.
Courtesy of Lillian Burgess

Dark blue clothing and summer straw hats mark this as an Amish auction, near Chesley, Ontario. Auctions mean many things in a rural community. Someone is ending their farming operation, retiring, or moving on to another part of the country and cannot take all their possessions with them. The day of the auction can be a stressful time for the owner but, for the community, it's not only a time to buy equipment but to meet the neighbours and have some homemade pie and coffee. This Amish auction meant that a good neighbour of mine, Joe Stutsman, was moving to Wisconsin, to another Amish community and I probably wouldn't see him again. That day I bought a spade and a threshing belt as well as three cups of coffee and as many pieces of pie, which cost me more than the tools I purchased!

The bank barn is a good standard 40-foot by 60-foot structure and has all the mixed-farm amenities — stalls for the horses, pens for the pigs and one or two stanchions for the milking cow. Since they don't use electricity, the Amish here need windmills for pumping water to the barn and house.

In the late nineteenth century silos became part of the farm operation to store silage for livestock. One of the earliest strategies was to construct round silos made from wood inside the barn to keep them out of the weather. Round was intrinsically the strongest structure and using hoops was a transferred technology from barrel making.

This barn with its interior silo is located in Port Elgin, Ontario. Built in the 1880s, the silo was constructed from long, narrow lengths of 2-inch thick tongue-and-groove spruce and bound together with iron hoops. The 12-foot diameter by 25-foot high silo served to store silage, usually a combination of green grains or green corn that would ferment into a nutritious, digestible feed. The barn and ramp have a wonderful pebble foundation, the stones carefully embedded into cement to add a bit of embellishment to the otherwise plain barn.

The rich brown of soybeans almost ready to harvest contrasts nicely with the red metal cladding. The graceful extended roof on the dormer protects the main doors of this barn near Lamlash, Ontario.

The stripped down four-bent timber frame.

The complex set of connections and intersections of a post and beam structure.

Take a barn down only to put it back up — that's the Mennonite way. Mennonite families moved to this farming region of Ontario, near Chesley, in 2002. Still driving horses and buggies for travel, they settle areas as a community, erecting churches and schools along with developing their farmsteads. Recently, a young Mennonite family purchased this farm, which came with a 120-year-old bank barn. They decided that the barn needed to be moved about 100 yards to a new location and an addition added for a dairy operation. The barn was stripped down to its original timber frame and then dismantled with the help of more than 75 Mennonite neighbours. The next day the frame was up again on its new foundation and, within a week, the new roof and siding were on. It shows the versatility of post and beam construction — remove the pegs that hold all the beams together and the barn can be taken down and used again.

Part of Mennonite barn renovations is installing a central vent: a wood or metal rectangle built in the center of the barn, which comes out through the roof in the form of a box with a protective lid. This natural thermosyphoning method of venting needs no electricity for fans.

Today many of the old farms in this region, originally mixed operations, have been converted to raising show horses because of the beautiful landscape and the proximity to Toronto, Canada's largest city. These barns near Rocklyn, Ontario, would have been built around 1880s, one to store straw and the other for hay and livestock.

There was a spring lushness to this region when I visited it but the land is poor and many a pioneer toiled here to little avail. I found this basic, unpainted barn in a valley near Brookville, Pennsylvania. A few barn boards are missing and the roof needs some attention, but the timber-frame structure still holds the barn straight and true.

Winter is a busy but peaceful time of the year for me: felling trees for next year's firewood; blowing out the driveway; skiing, and fixing all those things that broke during the hectic summer months. The area where I live, Sullivan Township, Grey County, Ontario, is rolling, partly wooded farm country, settled during the 1860s-1880s and still dotted with family-operated beef and dairy farms. All the old barns are timber framed, anywhere from four to eight bents, and commonly forty feet wide. This barn, just a mile from my farm, is typical of the area, being 40 feet by 60 feet and housing beef cows.

The sheer quantity of boards needed to cover this immense structure is astounding, but when the land around Cambridge Springs, Pennsylvania, was settled in the 1800s, there was no shortage of trees to cut for lumber!

I felt the raw energy of many years of farming emanating from this working barn. Massive and clean-lined, its original timber-frame structure, unencumbered by additions, measures a whopping 116 feet in length.

White-painted trim accentuates the windows of this early 1800s barn in French Creek, Pennsylvania. Upstairs windows in barns may indicate a chicken operation and windows downstairs often signify dairy stables, as in this case. A small, cement-block milk house, with the roof caved in, stands at the end of the barn. In 1753, George Washington changed the name of this town from Riviere Aux Boeufs to French Creek. This region played an important part in the French and Indian wars and the settlement of North West Pennsylvania.

Many barns have different types of advertising but this message is intended for the students at West Forest High School, located next door to the barn, near Tionesta, Pennsylvania.

There are beautiful fieldstone walls on this barn built by Irish-Scottish masons 200 years ago. Note the decorative S-shaped irons on the ends. A common construction feature on stone and brick barns, they act as washers for long iron rods that traverse the length of the barn to the opposite stone wall keeping the walls from pushing out.

I was up early on a clear October morning trying to beat traffic as I drove the narrow hilly paved roads southwest of Harrisburg, the capital of Pennsylvania. I happened upon this well-kept stone barn near Mount Zion in Cumberland County. Of the 10,000 barns built in that county, 1,000 stone barns remain standing.

Most were built between 1790 and 1840 by Scotch-Irish masons as banked barns. The two-story structures were built into the side of a hill, to allow for an extra floor below, providing stable room. The mow was then free for threshing and to store hay and grain for the prospering livestock trade.

In the second half of the nineteenth century, barns reflected not only the practical and necessary features of farming but decorative elements as well. These Victorian-style bank barns added ornaments to the roofs, doors and windows, in an effort to give a sense of distinction to the ever more prosperous farms of eastern USA. Practical as well, the louvers and cupolas on this barn near Arendtsville, Pennsylvania, provided air circulation.

In July of 1863 one of the greatest battles of the American civil war was fought at Gettysburg, Pennsylvania. Close to 175,000 men took part in the three-day battle and more than 50,000 were killed or wounded, the greatest number of casualties ever on North American soil in one battle. The fighting began about 8 a.m. on McPherson's Ridge just beyond the McPherson barn as Union cavalry confronted Confederate infantry advancing toward the town of Gettysburg along what is today Route 30. The fighting raged in the field between the fence and barn until late in the afternoon when the Confederates broke through the Union line at McPherson's Ridge. Amazingly, this beautiful stone barn with the white-walled wooden forebay survived, and is now part of the Gettysburg National Military Park.

This fine example of a typical Pennsylvania German bank barn stands on the western side of Carlisle, Pennsylvania. The two-story timber frame has the characteristic cantilevered forebay, an idea brought over by German and Swiss peoples in the nineteenth century. Imagine the huge timbers needed to support the forebay extension! Farmers tried to orient the forebay to the south so that cattle, which became an important agricultural product in the mid nineteenth century, had a place to shelter. Since there is no cupola, louvers located on all four sides kept the air circulating inside the barn to keep the hay dry.

A striking and well-kept Pennsylvania brick barn with two extended brick additions near Fayetteville, Pennsylvania.

The brick extensions were used as granaries and guarded well against mice. The beautiful brick patterns of the barn, used for venting, break up the blank face of the large gable walls. Each shape has a different a meaning. On this barn the patterns show a sheaf of wheat on top, a wine glass in the middle, and a haystack on the bottom.

The barn is situated beside a busy highway and will soon be converted as part of a development that includes a golf course. Although the historic barn already looks removed from its agricultural heritage, I am pleased that at least it's being preserved for the future.

With the aptly named Blue Mountains as the background, this farm near Mount Holly Springs, Pennsylvania, has been a working operation for at least 200 years. Both the barn and shed are made from brick with diamond "haystack" patterns.

Just south of the interstate near Carlisle, Pennsylvania, at a busy intersection stands this brick-end barn. With additions on every side, this is the only part left that lets you see the wonderful ventilator patterns which include this "unfolding lily".

Vermont Highway 100 winds around mountains and streams all the way from north to south and along this historic route are many wonderful barns. This one, owned by Patricia Sears and Steve Mason, was built in 1865, south of Westfield. It is a six-bent, timber-frame structure measuring a standard 40 feet by 60 feet. All the bent girts (horizontal timbers) are remarkable single 10-inch by 10-inch 40-foot pieces of white pine, while the purlin plates are 8-inch by 8-inch 60-foot one-piece timbers! It's amazing to picture the size of trees available 140 years ago. The structure itself is an anomaly.

Posts usually reach to the purlins as one piece with the bent girts making an H. In this frame, the posts stop at the bent girts and continue again on top, connecting in this way with the purlins. Unusual as well, is a collar tie that connects the two posts at the top of the roof, making a box-like shape. Although strong, it leaves too little room along the top of the inside ridge for a hayfork, a timesaving device that most farmers installed in the early 1900s, after this barn was built.

After many additions, including a modern silo, only a small amount of stonework remains visible on this Mennonite farm near Shippensburg, Pennsylvania.

Stone all the way to the top of the roof on the gable end! Imagine lifting each heavy stone up the high wooden scaffolding to place among hundreds of others as the wall slowly went up. The very narrow slits indicate an older barn, probably more than 200 years old. The forebay walls were built of wood because the sleepers (joists) could support a lighter timber frame — but not cantilevered stone walls — and the tons of hay inside. Near Boiling Springs, Pennsylvania.

When a barn's bottom timber plate, which rests on top of the foundation, starts to sag because of improper support, the roof and siding follow. Originally, there was a forebay, now partially covered over, in the front of this barn. Unfortunately, there was never a proper foundation built and over the years the barn began to take the shape of the hill it was resting on, near Jeffersonville, Vermont.

Big, open cash-crop fields are prevalent in this flat part of Ohio near Urbana. The rectangular section of this barn was built in 1875 and the octagonal structure added in 1897. A new foundation was put in recently by jacking up the barn and pouring concrete underneath. The slate roof is original but Michael Sower, the present owner, needed some slate shingles since a few broke over the last 100 years. Luckily, a barn across the road that was to be demolished had the identical slate roofing, which Michael was able to obtain for fixing his roof.

Round and Polygonal Barns

ROUND BARNS — INTRODUCTION

LIKE OTHER BARNS, round and polygonal barns were built to house livestock and store feed and grain. But they incorporated very different architectural strategies. The visual impact of the round barns' domes with their radiating rafters brings a pleasurable shock, something grand, something unexpected for a building that is, after all, used to house livestock and the manure and dirt associated with them.

I experienced that amazement when I saw my first round barn near Evart, Michigan. The country there is flat, a little swampy with few farms. The Feikemas' 1907 round barn came into my vision, as foreign as a flying saucer settled on the ground but ready to lift at any moment. I pulled into the driveway and Fred and Ferne, the owners, came out of the house and graciously gave me a tour, pointing out the vagaries of the barn. Both retired from farming, but still very enthusiastic about their barn, they pointed out the tall inner ring of round posts inside, each more than 25 feet in length, reaching for the roof that radiated rafters from a center ring. "Built with farmer ingenuity," said Fred, wearing a cap that had a local farmer's co-op trademark on it as he shuffled along.

Round and polygonal barns show an amazing architectural concept and skill of design in their construction. It's almost as if the circle, the perfect shape, inspired farmers

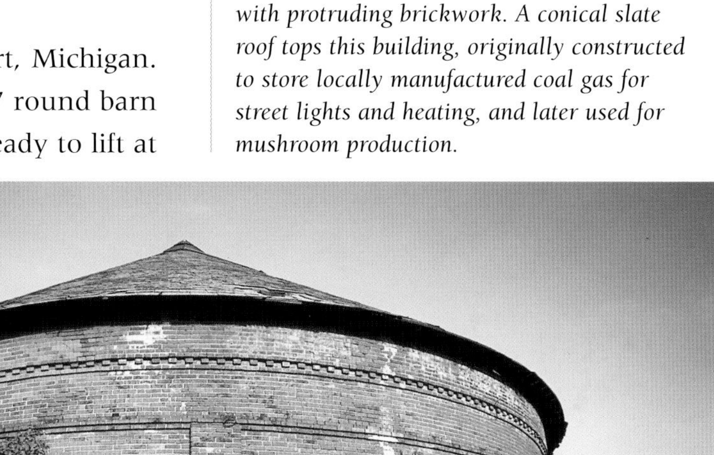

Built in 1889 and now part of Oberlin, Ohio, this round building was constructed with brick pilasters or columns every 10 feet for extra strength while the walls were decorated with protruding brickwork. A conical slate roof tops this building, originally constructed to store locally manufactured coal gas for street lights and heating, and later used for mushroom production.

The structural supports for the round roof became graceful, each piece of lumber a small part of a greater framework that holds up a giant roof. Trusting mathematics, the builder would not know how the roof performed until it was complete and snow accumulated to test the integrity of the structure. Many vents and windows, matching in size, add to the overall effect of gently balanced symmetry.

with an interest in math. It was also said that the devil could not hide in the corners of round barns.

The initial experiment in round or barrel barns came from the Shakers, a religious group in Massachusetts who built a 270-foot-diameter barn in 1826. But that is not to say that all the farmers who built round barns after that went down to have a look!

It wasn't until the 1880s that round barns became popular and most in North America were built during that era. By then state agricultural boards and agricultural colleges had begun touting progressive farming methods based on efficiency and cost-effectiveness. Round and polygonal barns exemplified the progressiveness of the new technological age. Experts wrote in farm magazines that round barns would require fewer materials and produce more efficiency of space than conventional rectangular barns. Because of the self-supporting dome roof, the interior was free of cumbersome posts and poles. Livestock on the ground floor could be stationed in stalls in a circle, taking up less room. The farmer could do the chores moving around the circle in one direction, making it easier to feed the livestock and clean the stalls, without having to go back and forth as in a regular barn. And those round barns with a circular wagon drive on the third floor allowed hay to be unloaded down into the mow below, a great saving of energy.

Another advantage extolled about these barns was their wonderful natural light produced by the central cupola as well as windows set in the face of the circumference. Ventilation was also more efficient, with air flushing through the middle of the barn through the cupola. By the end of the nineteenth century, as silos became more accepted, they were incorporated into the middle of the round or polygonal barn, often supporting the cupola or the roof structure.

Silage was easier to feed out that way since it was central to the stables. As well, the aerodynamic shape of the dome roof and barn minimized the damaging effect of storms.

But farmers are a conservative, practical lot, and they like to see new farming methods tried out before they attempt anything. So, farmers finally began to build round barns because . . . another farmer down the road who was a little bolder had constructed one. In Quebec, dozens of round barns were built around 1900, all within a 100-mile area. The first ones were constructed because across the border in Vermont, other farmers had been successful at it.

However, this design had some disadvantages. It definitely required building skills, so that often, costly carpenters would need to be hired. Lumber used for the outside finish had to be cut more often and in shorter and narrower pieces. If clapboard siding was used, narrower segments were needed than for a flat wall, and the barn had to have a large enough circumference that sections of clapboard could be bent to fit after being soaked. Another disadvantage was the challenge of additions. New buildings were rectangular in shape and therefore hard to fit onto a circle.

Finally, the technological twentieth century spelled the end of round barns, as it did for timber-framed ones. Farming became more industrialized and specialized, and materials became cheaper while labor became more expensive, making both round and timber-framed barns too labor-intensive and therefore too costly to build in this new efficient age.

Not too large, with plenty of room and in decent shape, this barn is a hobby farmer's dream. Just off a main highway on some hilly ground, near Ferrisburgh, Vermont, this 8-sided barn, built at the beginning of the twentieth century, looks perfect for a few horses or some goats with enough storage capacity for hay in its mow.

This beautifully symmetrical structure has a cut stone foundation built into the hill to make room for stables below. A driving shed on the right was used to house prisoners who helped on the farm during the Second World War.

Michigan Round Barn

It was a railroad man's dream: buy a farm and build a barn designed like a railroad round house. Today that barn is still standing, albeit leaning a little, a testimony to a dream come true.

Built in 1907 near the town of Evart, Michigan, the 60-foot diameter structure was constructed from trees cut on the farm. Inside are two rings of posts. The outer ring, which supports the outer edge of the roof, has a cedar post every two feet for maximum strength. About 16 feet in, another ring of posts, seven in total and about 25 feet in length, braces the hip of the roof. On them a circle of bent wood supports the hip structure. Although timber-frame notching techniques have not been used, and the joinery is rather crude, the barn still stands an amazing 56 feet to the peak. Around that inner ring runs a circular track for the hayfork, something unique and probably borrowed from the railroad dream.

This structure was inspired by a railroad man's dream: buy a farm and build a barn based on the railroad round house. Today this barn still stands, albeit leaning a little, a testimony to dreams that do come true.

Part of the cut stone foundation and small round window near the bottom show a typical Michigan welcome for visitors.

In 1928 a tornado ripped off the whole roof, including the cupola. A slightly lower-pitched roof — without the cupola — replaced it. During the Second World War, a driving shed on the farm housed prisoners who helped with the farm work. Later the same shed housed newly arrived immigrants. When Fred and Ferne Feikema bought the farm in 1969 and found only cardboard for insulation in the shed, Ferne wondered how those people survived the cold winters in mid-Michigan.

The roof looks a little wavy because Fred replaced the aging shingles with spray-on foam and then painted it white. Fred told me that this worked well originally but over time woodpeckers have created some leaks when they made holes in the foam for their nests.

This 60-foot diameter round barn in Michigan was built from poles set into the ground, a much simpler process than the more common timber frame, but it still stands strong today.

A devastating tornado in 1928 ripped the roof completely off. Although the roof was rebuilt, the cupola was never restored.
Courtesy of Fred and Ferne Feikema

The inside view of this Michigan round barn's roof is a wonder of architectural complexity though built by local workers with only practical experience.

Ferne remembers when they were farming beef cattle. Each fall the barn would be full of square bales, all ready to feed the livestock during winter. "I would climb to the very top of the eaves, on top of the bales, to put plastic over the windows up there. That's how much hay we used to put away," she said.

The Feikemas are proud of their round barn. "It's the only round barn in Osceola County," said Ferne. She had all the information about the barn at her fingertips as she had recited it to so many visitors who have stopped over the years to look and take pictures. "We're gonna try and keep it up, but it takes a lot of money," said Ferne.

A local university has offered to help but only if they can take down the barn and move it to their city. "We've said no," said Ferne. She and Fred would rather the barn stay in Osceola County.

Ontario Round Barn

Thomas Tweed Higginson was completely intrigued by the Chicago World's Columbian Exposition held in 1893. He admired the fourteen grand buildings of the Fair which were all in the "Beaux Arts" style, with conical roofs and arched entranceways. In the 400,000-square-foot Agricultural Building, he saw model farm buildings, including round barns. And, for 50 cents, visitors could ride the Ferris Wheel, just being introduced to the public and the Big Cheese, another ride in the shape of a round block of Swiss cheese.

Thomas became so inspired by what he had seen that he decided to build a round barn at his Hawkesbury, Ontario farm. He made a scale model of his proposed barn and took it to show his local Agricultural Society. They laughed and said it couldn't be done.

In the spring of 1893, at the age of 64, he began the construction of his barn. We have unusual insight into the whole project because he had kept a daily journal of his activities since he was 15 years old. His great-granddaughter, Frankie Higginson, still possesses his journals.

The central silo is made from spruce. Tall posts support the intricate bottom of the roof.

I visited with her on a sunny fall day. She told me many stories about her great-grandfather and was kind enough to let me look through his journals. Already a grandmother herself, she was enthusiastic about showing them to me. Some were small, with tiny entries and others larger, with bolder, simple writing.

After putting in the 76-foot diameter stone foundation and some of the inner pillars, Thomas wrote on June 18, 1893, "It looks like the Big Cheese of the World's Fair." Two months later he described the on-going construction as a "circus ring," another of the prominent events he had seen in Chicago.

Slowly his dream came true. By October 1894, he wrote that his brother John had "nailed the last [cedar] shingle on the roof." All who helped wrote their names on the underside of the shingles. One hundred years later Frankie, who still lives on the farm, was burning some of those shingles in her stove, as they had been replaced by steel roofing. She found some that had been signed. "I saved them all and am still looking for any more," Frankie told me.

The barn construction uses 20-foot, 6-inch by 6-inch posts every three feet to form the outside wall on top of the foundation. About eighteen feet in, another ring of longer posts supports the conical roof about half-way up. Interestingly, the top of the inside silo supports the middle of the roof and a large cupola, which ventilates the mow and silo, adds some light. The roof is made from four-inch by four-inch posts every two feet. To give the whole structure integral support, horizontal boards cover both roof and outside walls. Wooden shingles were the original roof covers while board and batten protected the walls. The stables had rings of stalls for horses, cows, pigs and chickens around the central silo and outside wall. There are no longer animals in the barn.

Other journal entries mention farmers coming from as far away as Chicoutimi, Quebec, to see how Thomas had built his barn. Some took his ideas back with them and built round barns, perhaps influencing the construction of a dozen or so round barns built in the Eastern Townships of Quebec at the beginning of the 1900s.

A typical farmer, Thomas started every journal entry with the weather, then listed the size of the crops he harvested. An entry from the fall of 1895 states "125 bushels of potatoes and 12 tons of clover" were taken, "a good crop." Corn fodder, stored in silos, was a new and important crop. The middle of his round barn featured a round wooden silo. Bent strips of spruce nailed on the outside to squared posts covered an inside made from bricks. Plaster on the bricks gave a smooth finish so the corn fodder would not stick to the wall.

As far back as 1875, Thomas listed the names of horses, bulls, and cows and the number of sheep and hogs. Sixty hens then were worth six dollars, or ten cents a chicken. A big bay horse called Lady Dufferin sold for $125.

A schematic drawing in Thomas' journal, drawn before he built his round barn, showed the size of each field and the placement of buildings. The sketch shows a barn located just beside the present-day round barn. It was drawn as a cross, or a barn with two additions on each side. What happened to that building Frankie doesn't know but, with the building of his round barn, Thomas probably just didn't need it, or didn't think it was good enough.

Inspired by the Chicago World's Columbian Exposition of 1893, this 73-foot diameter round barn has a simple gracefulness. Farmers from hundreds of miles away came to see it, taking back with them visions of building their own round barns.

This working round barn has a center wooden silo built inside while the barn was being constructed in 1918, hence the large cupola acting as an air vent on top of the domed roof. Located near Kingsley, Michigan, this barn also has guy wires strung from the ground to the structure to keep it from leaning any farther.

One could imagine a huge can opener prying open the dome roof from this large round barn near Palmyra, Michigan. Unfortunately, the barn has not been maintained properly and rain and wind will soon have it completely torn down.

A very unusual brick round barn built west of Greene, New York, in the Chenango River valley. Brook Haven, printed out in clay tiles on the front of the barn, was built during the Great Depression in 1931 and was quite fanciful for its time. The brickwork is especially decorative, and stone lintels grace many windows on the second story as well as the stables below.

This round barn stands in the quiet of woods and yards. Deer were feeding on the grass the August day I visited. This lovely example of the best of artistic nineteenth-century barn building is rich in brick detailing with round enhanced louvers and curved windows. It is approximately 100 feet in diameter and located within the boundaries of Urbana, Ohio.

The Manchester barn presents one of the finest examples of round barns in the USA. Based on the Shaker design, it was built near the town of New Hampshire, Auglaize County, Ohio, in 1908 by Horace Duncan. It measures an astounding 102 feet in diameter. Hay and feed were stored in the center and livestock was housed around the circumference. The farm has served the Manchester family for five generations.

I had been admiring a round barn near Barnston, Quebec, with my friend Francis Loomis, a fellow barn lover, when the owner told us there was another just up the road. Now Francis, who had lived in the area all his life, had never heard about this barn and we were both excited to see it. The directions were typical of a farmer's: Take a certain little road up over the hill and we couldn't miss it. The road disintegrated into a long, uphill dirt track and, just as I was fretting about possibly being on the wrong road, we emerged into a clearing of broad green lawns, a beautiful renovated house, and a freshly painted and restored large red barn. A hidden round barn, what a thrilling find!

This stud-framed 70-foot diameter round barn has a circular third floor. It allows hay to be dropped into the stables below.

The "barn bridge" as it is called locally, leading up to the third story, was set upon flat stones, without any mortar between, and is still in wonderful condition. The present owner D. Sheard, found this old homestead abandoned, and over many years lovingly restored all the buildings, in the process preserving a bit of significant history.

Near Cambridge Springs, Pennsylvania, I was driving down a little gravel road, tired, and not looking too hard when this amazing white octagonal barn popped up, and with it my spirits. Unfortunately, no one was home but I found out from a neighbour that it was built in 1900. The 8-sided structure has the timeless look of the stately octagonal barns of that period. The grounds and buildings are beautifully kept, set off by the immense maple tree in front, which almost looks planned by a landscaper 100 years ago.

POLYGONAL BARNS — INTRODUCTION

Polygonal barns, featuring 5 to 16 sides, with occasional 20-sided ones found in the mid-west USA, became the vogue by the mid-nineteenth century, although the first and most famous 16-sided barn was built by George Washington at his home in Virginia in 1793. Polygonals were cheaper versions of the round barn, though the more sides they had, the harder they were to distinguish from the round. Earlier versions were wood sided, while later ones were built from brick or clay tiles. The multi-sided barn, the most popular being the octagon, was cheaper to build than the round barn because longer pieces of straight wood could be used, so there wasn't as much cutting needed, saving on labor costs. Although geometry still played a large part, measurements were easier to calculate. The straight, if short, sides made additions less complicated.

Like round barns, polygonal barns had central ventilation and silos, livestock that faced inward in a circle and labor-saving ways of handling hay such as the third-floor wagon drives from which farmers could dump hay onto the mow below. Most did not have freestanding roofs; inner central posts supported the many-sided roof.

Practicality still played a part in the decision to build a polygonal shape rather than the conventional rectangular one. Like the round barn, polygonal designs were promoted as having more room because of their higher volume-to-surface ratio than the rectangular barn. At first these barns were built with heavy timbers using traditional mortise and tenon joinery, but by the 1880s this was a costly venture both because of the lack of large trees

The classic polygonal barn with a small rectangular addition on one side.

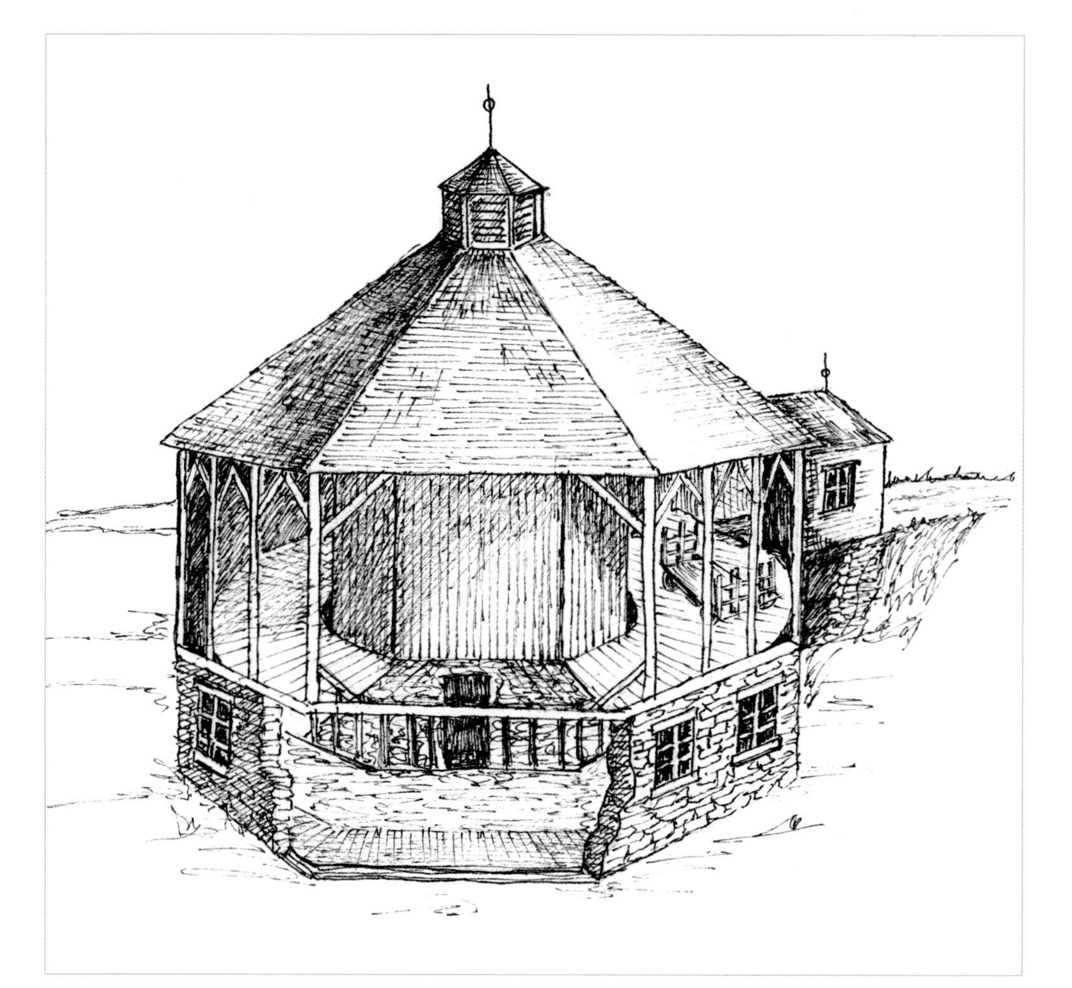

available and labor costs having gone up. Using poles and sawmill lumber became popular and cheaper, and many were built quickly this way, but without as much structural strength as the timber-framed types.

By the 1890s the multi-sided barn's popularity was waning and round became more popular but by the early twentieth century, both forms had become antiquated, as they did not adapt well to new agricultural technical advances.

Although round and polygonal barns would seem to have outlived their usefulness, I was surprised to see how many still house dairy cows, beef cattle, or sheep. Many have become county landmarks. An example is the Thumb Octagon Barn in Michigan, which was saved by local barn enthusiasts. This majestic 102-foot-diameter barn has been turned into a museum, showing off not only the barn itself, but also historic agricultural equipment and methods. Unfortunately, others not being used for farming or local museums are falling prey to the weather and lack of maintenance, and are quietly crumbling into the earth. They deserve preserving, for we are unlikely to see such unique agricultural buildings again.

A beautifully renovated, tall 12-sided barn which was built next to the rushing Mill Brook probably in the late nineteenth century. It is now an exclusive inn and restaurant in the mountainous region of Waitsfield, Vermont.

Michigan Thumb Barn

It's a sparsely populated region of Michigan, because farmland is prime and the fields go on forever. The spring day was sunny for a change, the fields were full of grain and corn, and I was looking forward to seeing the Thumb Octagon Barn, named for Michigan's distinctive land form.

Seasoned builders George and John Munro of Gagetown consulted with the local mathematics teacher to help them work out the calculations necessary to construct this intricate, three-story octagonal building.

In 1923 James Purdy, president of the Gagetown Savings Bank, hired the Munro brothers to build his "air castle", as his wife Cora described the unique building in her diary. It wasn't completed until 1924 and what a magnificent structure it was! The barn was featured in the June 1925 issue of *American Builder*. A quote from the article enthused, "The most striking feature of this group of [octagonal] buildings is this novel barn."

Why did James Purdy build an 8-sided barn? During this period, agricultural colleges were still promoting octagon barns as the buildings of the future. Agricultural leaders felt that it was more efficient to work in this shape of building than in the conventional, rectangular type and that the added space could save having to build outbuildings to house pigs, chickens, and other livestock.

This complex set of dormers, second-story cupola, and ventilators tops the Thumb Barn.

By the 1980s and '90s the Thumb Barn was almost in ruins, with large gaping holes in its roof and most of its hundreds of window panes broken.
Courtesy of Friends of the Thumb Octagon Barn

A bank owner in the little town of Gagetown, Michigan, hired local builders to construct this complicated octagon structure in 1923.

Just the sheer size of the Thumb Barn is imposing. Each of the eight walls measures 42 feet, 6 inches long. The diameter of the whole building is an impressive 102 feet. The tip of the weathervane stands 70 feet high, about as tall as an 8-story building. There are 32 windows in the upper level with 288 individual windowpanes and 8 shed dormers, the latter very unusual for a barn, each containing its own windows. Since Purdy was afraid of fire burning down his dream, each window was positioned to prevent direct sunlight shining on the hay and straw stored inside. Irregularities in the glass of that era could magnify the sun's rays and start a flame. The combined floor space, including the partial second mow/loft level totals 14,300 square feet.

Once used to pick up and move loose hay and straw from the wagons to the mow/loft area, a circular hayfork still hangs from the ceiling. Opening the main doors ensures a good breeze with air flowing up through the central shaft and out through a steel ventilator at the top of the roof. Purdy also built an innovative air venting system, constructing box-like air shafts that ran down from the roof to the second story and expelled air through the same ventilator. This was a practical way to disperse any heat created by the hay if it was damp, preventing a disastrous fire.

After Purdy sold his farm estate in 1942, the property changed hands many times and, after years of neglect, the barn was slated for demolition in the 1990s. Parts of the roof had caved in, barn boards were missing, and most of the windows were broken. In 1991 the estate ended up in the hands of the local bank, which sold it to the Michigan Department of Natural Resources (DNR) who were interested only in the 80 acres of land as a game reserve. Through the efforts of local residents the barn and house were saved and today Friends of the Thumb Octagon Barn have restored both buildings almost to their original condition and have preserved them as a historical landmark and museum.

You have a visceral sense of the enormity of the Thumb Barn when faced with this amount of wood and dwarfed by its height. The box-like columns are air vents that would suck away any hot air build-up in the loose hay stored on the second floor.

Ohio Freeport 16-Sided Barn

Freeport, Ohio, was settled by "foreigners", it says in the local history book but these foreigners turned out to be "Yankee Quakers."

This very hilly backwoods part of Appalachia is rich in natural resources: fertile land, hardwood forests (oak, hickory and beech), and coal; more recently, natural gas has been discovered. Farms in the 1840s produced buckwheat, oats, wheat, sorghum, apples, maple sugar, and many kinds of vegetables, from which pioneers scratched out a living, trading any excess with one another or in local towns.

About a mile from Freeport, along the twisty Skull Valley Fork Road, John B. Stewart decided to build a barn like no other in that region, for it was to have sixteen sides. The present owner, Oliver Workley, told me John was a finicky man. As a cabinetmaker he once built some oak shelves that warped after the customer received them. Instead of fixing the shelves, John burned them and made new ones saying that he was a "perfectionist".

The owner who built this intricate roof was a cabinetmaker and it shows in the fine details of construction. A barn like no other was built in this region of Ohio by John Stewart in 1918.

He applied the same standards to his 60-foot-diameter barn, constructed during 1917-18. The structure was made from oak and black walnut and meant to last more than 100 years. Each of the 16 walls is 12 feet wide and 12 feet high and boarded diagonally on the outside for extra strength. He built the walls first and then, inside, the silo, which was as high as the roof of the barn, with the cupola on top for ventilation.

I sat on the front porch of a hillside house on a warm summer's afternoon, overlooking the barn with Oliver while he proudly showed me an old photo of the construction of the barn. The silo is built and surrounded by wooden scaffolding and the walls are planked. There is John's son, still dressed partly in his First World War uniform, helping his father build the barn, "I was told when he came back from the war there were no clothes available to buy so he just wore what he had," said Oliver.

The wood silo could hold 100 tons of silage corn, but John filled it to the top only once — that corn lasted him five years.

The carefully laid foundation was cut from sandstone, quarried from the 100-acre farm. At each of the 16 corners, the sandstone rocks were cut at the appropriate angle.

John milked cows, kept pigs and horses in the stables plus as a specialty, Angora goats for wool. He installed a time-saving "manure caddy," as it is called in these parts, to carry the manure from the animals out of the barn. The single-rail carrier went all the way around the stables and out the bottom doors to a manure pit. From there manure was loaded onto a spreader and taken to the land.

The roof is a complicated piece of engineering, the top part an 8/12 pitch while the bottom, steeper section is 21/12. Its cedar shakes lasted more than 50 years until Oliver Workley bought it in the early 1970s and roofed it with asphalt shingles. Oliver told me he bought the farm from a lumber company. The barn had fallen into disrepair and he "saved it just in time."

The 16-sided Freeport, Ohio barn during construction in 1917 and 1918. Boards were put on diagonally for extra strength and the silo was made of local oak. The owner's son had just come back from the Great War and is still wearing part of his uniform while helping his father, John Stewart.

The wooden silo was used for corn silage but because it was so huge it was filled to capacity only once.

Pennsylvania Octagon Brick Barn

This grand octagon brick barn, built as a showplace, has served as a local landmark since it was constructed in 1879 near North East, Pennsylvania, just a mile from Lake Erie.

Construction of the imposing 8-sided building, measuring just under 100 feet in diameter within the interior walls, was accomplished by brawn, block and tackle and horse power, even when it came to lifting the great beams for the roof structure. From the center of the stables to the peak of the roof is almost 60 feet. Until a violent storm in 1957 ripped the cupola off the center of the roof, the building stood even taller. This roof structure, which has no central supports, is woven together by a series of small triangles, (trusses in today's terms), all pinned together by mortise and tenon joints.

Each of the eight walls has three layers of brick with air spaces between acting as insulation. "In summer it's always cool in the stables and in winter it rarely goes below freezing," said John Phillips, Jr., present owner and vineyard operator. Phillips Vineyards uses an image of the stately octagon brick barn as a logo on its business cards and other promotional material because the barn is such a landmark.

At the back of the barn where the ramp and slight hill meet the structure, the foundation is made from two-foot thick, beautifully laid stone. At the front, the foundation is hidden but the brick wall mounts an impressive 42 feet to the eaves. Added columns of brick or stone reinforce every corner, giving the barn somewhat the air of a small medieval castle.

The walls of this well-kept octagon barn, built in 1879 near North East, Pennsylvania, are a solid three bricks thick.

Because the land is so close to Lake Erie, enjoying a stable and moderate climate, the farm has always produced a variety of fruits. Presently the operation grows 120 acres of Concord grapes, used primarily for Welch's Grape Juice. During the late 1950s and 1960s, John Phillips, Sr., who acquired the octagon barn and 80 acres around it in 1955, had more than 400 acres of fruit under cultivation, including sweet and sour cherries, apples, peaches, plums, and grapes. He converted a local barn to house the more than 200 laborers they needed seasonally to harvest that much fruit.

In the 1970s the barn was used as a theater, a venue for local plays. During one of the state elections in the 1990s the barn and beautiful lawns were utilized to showcase the Republican candidate running for governor.

Again a working barn, the bottom floor is crammed with vineyard equipment and upstairs, what used to be the hay mow stores a wide variety of farm machinery.

Standing inside the building, looking up at the graceful roof, I marvel at its intricacy and strength, all built with nineteenth-century manual craftsmanship.

A skilled and attractive transition from a brick to a stone foundation.

The beautiful arched main doorway.

Inside the Mystic barn on the rotating platform. The great double doors opened onto 12 bays, each one used for a specific purpose.

Quebec Mystic Barn

Frances Walbridge is a tiny woman, now in her nineties, with a lively mind and sparkling blue eyes. She spent many years as a missionary in the African country of Angola, but returned to Mystic, Quebec, to care for her aging father. One of her earliest memories is of coming as a young girl to her grandfather's farm in the Eastern Townships during Easter and looking across his man-made lake to islands planted with circles of white, fragrant narcissus.

Her grandfather, Alexander Walbridge, was an extraordinary man. In the quiet rural village of Mystic, he built up a successful foundry business, inventing a diverse line of industrial and agricultural machinery. With some of his profits he constructed an ostentatious family home on the farm, a mansion which included a conservatory where he grew bananas and an aquarium stocked with alligators. In front lay his 10-acre lake with islands and boathouses and behind, an incredible 12-sided barn.

Built in 1882, the polygonal barn borrowed details from a railroad roundhouse. What was really ingenious about this timber-framed building, besides the complicated architecture, was a cast-iron turntable manufactured at Walbridge's iron works foundry and set in the center under the cupola. The idea was a practical one because it allowed avoiding the difficulty of backing a team of horses once their load was lifted off. In this barn, horses and wagon entered through the main entrance and moved onto the cast-iron turntable. By means of water-powered gearing, the whole platform rotated like a gramophone record, more slowly of course, until the team reached one of the 12 bays. There the wagon could be unloaded by a hayfork, also water-powered and then the turntable was rotated again until the horses could walk forward out of the barn.

This incredible barn with 12 roofs is 78 feet in diameter and formerly housed a 30-foot turntable. All were supported by posts in the stables forming two concentric rings with a stone foundation forming the perimeter. Courtesy of The Mystic Barn Conservancy

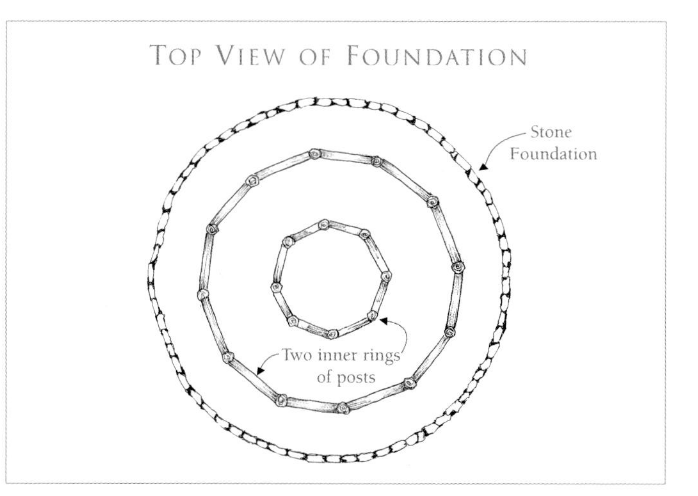

Another view of the Mystic Barn with Lakelet Hall in the background. The mill, the two-story building on the left, powered the turntable inside the barn by a long belt, much like a belt pulley on a tractor.
Courtesy of The Mystic Barn Conservancy

Alexander Walbridge was called a visionary. Among other things he built this 25-room "English Manor" using bricks he manufactured from clay dug on his farm. Inside was everything an upper-class Victorian family would want including a conservatory growing banana trees and an indoor pool holding alligators. Named Lakelet Hall, the house was joined to the barn by a double brick wall built to decrease the chance of fire spreading. The beautiful mansion was torn down in 1941 because its upkeep became too costly.
Courtesy of The Mystic Barn Conservancy

Walking into this barn is like waking up in the laboratory of an early nineteenth-century scientist, the one who was going to build a fantastic flying ship from wood, cloth and iron. Looking up through the filtered light to the central cupola, you see suspended the large cast-iron gearing that once operated the turntable. The roof's timber truss supports soar like wings, all ethereally illuminated by small windows that surround the cupola, forming a scene in one of H.G. Wells's books. Just below the cupola, but still well above your head, the twelve great double doors, most hinged about halfway up, with diamond shaped windows in each bay, give the impression of an otherworldly place.

All went well for Alexander during his heyday in the late 1800s. Eventually one of his biggest customers, South Eastern, contributed to putting him out of business. The railroad defaulted on its payments and Walbridge challenged the company to a lengthy and costly court battle, which he lost.

Financially crippled and faced with tough competition from larger foundries, he was forced to close his business.

Not long afterward, in 1897, Alexander was doing chores in the stables of his polygonal barn. As he climbed the ladder from the stables to the main floor, the trap door fell on his head. A few days later he died of his injuries at the age of 69.

Walbridge's family managed as best they could but in time they could no longer afford to live in the great 25-room mansion. They moved back to the property's pioneer farmhouse which had been used at one time for the servants and field hands. The mansion fell into disrepair and was demolished and the material sold for salvage in 1941. A flood destroyed the dam with the waterwheel that once powered the turntable in the barn and subsequently the man-made lake with the islands full of beautiful flowers dried up. During the 1960s the cast-iron turntable was dismantled and used to replace a local bridge but that later collapsed and the material was scrapped.

Today, the original 1843 farmhouse, still home to Frances, remains as does the great barn. In front of the spectacular red-walled structure that attracts many visitors each year, stand two large oaks, which Alexander Walbridge planted more than 100 years ago. The farm and its buildings have been placed under the protection of the Walbridge Conservation Area Limited, mainly to preserve the barn and finance repairs. A request has been submitted to the Quebec government for official heritage status. If ever a barn deserved to be preserved and admired for its historical significance, this unique 12-sided structure should qualify.

Looking up at the inside of the cupola, with much of the plaster having fallen off, one can see the gearing once powered by a small creek just behind the barn. The mechanism rotated the turntable and moved the hayfork.

In the stables, massive timbers on cement pillars formed the inner circle that helped support the floor above.

Frances Walbridge, one of three sisters who still reside in Mystic, Quebec, granddaughter of Alexander Walbridge, holds one of the chisels her grandfather manufactured in his foundry and then used to build the 12-sided timber-framed barn.

Even with a local map I got lost in this picturesque but rough country of Vermont, near Ludlow. The little pockets of farmland have led some entrepreneurs to raise livestock for a niche market allowing them to stay in this wild hill country. I learned from my long-time friends who live nearby that the owners of this recently built 8-sided barn and its many outbuildings raise sheep and sell their organic lamb and mutton cuts in New York City.

Each wall of this 8-sided barn measures about 27 feet, with wooden louvered vents placed on seven sides, the exception being the wall with the main doors facing the roadside near Summerville, Pennsylvania.

It is such beautiful countryside in this part of Ohio, with its hills, small winding roads and many historic barns. I was anticipating greatly seeing this octagonal barn but disappointed at its condition because at one time it was a noble barn. The farmer painted it a brilliant white with red trim around the windows and doors to accent his showpiece building.

Fortunately, the 60-foot diameter on-grade structure has been saved from completely falling into the ground by the local Harrison County Historical Society. They plan to move the unique barn from its present location, near Moorefield, to the center of the county on Highway 250. There, it will be renovated to its original condition and turned into an agricultural museum. Such a prize 8-sided barn, with its arched louvered windows and ingenious timber frame roof support, should be saved for its historical value

A cement-block octagon barn is a rare sight but there is good reason for this one, built in 1914 near Edmore, Michigan. The farmer who designed it was also a cement contractor called F.W. Johnson & Son. His company had the forms for cement blocks, which they used to build this unusual barn. Johnson also constructed the sidewalks in nearby Edmore and some of the local cement bridges. Also known as the "Potato King of Montcalm County," he was a "master farmer", according to Lee Peterson, who bought the farm from Johnson in 1958. An innovative farmer, Johnson bought and used an airplane to dust his potato crops as well as using the wind from the airplane's passage to keep frost from settling on the crop on cold mornings. Unfortunately during the 1950s Johnson lost one whole crop to frost when his son, the pilot, didn't get up in time.

The attached red gambrel-roofed barn, built in 1940, has a nice hay hood to protect the hayfork used to pick up the hay and straw from wagons parked on the ground below. The fork would then travel along its track inside the barn and drop the hay or straw, which was used for feeding or bedding livestock during the long, cold winter months.

I spotted this huge octagon barn miles away as it rose out of the flat fields of Michigan like some building module out of the future. Exceptionally tall, it stands a whopping 92 feet at the peak because of the high basement walls, which add 20 feet below the main floor. Each of the eight sides is 38 feet long, giving a very roomy interior for equipment, hay and feed storage. The building, probably dating from the early twentieth century, is located near Elsie. No longer in use, it still looks in great shape and I hope it will be used again for farming or at least for heritage education.

I was driving through a beautiful area of New York State, hilly, forested, dotted with farms, looking forward to seeing this early example of a 12-sided barn in Jefferson, an area that was settled in 1803. Just as I was leaving the village, I spied the barn on my right, while heading up a hill on a busy highway without any shoulders. I pulled off as much as I could and took a few photos, but I had many irate motorists glare at me, as they waited for opposing traffic to pass. I love the matching 12-sided cupola, a nice touch to this stately building.

I was driving down a little paved road on a sunny day in rural New Hampshire, just coming out of Plainfield. I wasn't expecting to see any barns yet, but out popped this barn-like structure. It turned out to be the Plainfield Christ Community Church. Newly built, it incorporates many barn motifs including the main red barn structure, a wooden silo, cupolas, barn-board siding and a weathervane in the shape of a fish.

Variations

INTRODUCTION

A NEIGHBOR OF MINE who has been farming all his life told me once that there are no two barns alike. And it's true. For every type I have listed, there are still no two English, Bank or Dutch barns exactly alike, as every farmer has built his or her barn into a landscape and each has a farming operation different from the neighbor's down the road. Types offer a general pointer for an era that used ethnically and environmentally based building methods to produce barns that were similar but did not have the cookie-cutter exactness of later factory-built types.

Even during the height of timber-frame barn building, there were some that fit no category, such as the majestic Star Barn and grand Enfield barn in this chapter.

As the nineteenth century came to a close, building methods gradually changed. The gambrel-roofed barn became the standard, placing a roof style found in colonial New York State homes over the body of the English barn. Traditional timber framers began to adopt newer methods to their designs, incorporating the gambrel roof as part of their construction techniques. The double-pitched or hip roof became popular because it used shorter rafters, at the beginning still made from half-round logs, but quickly changing to planks and dimensional lumber. Amazingly, even though we think of traditional

A large barn being built in 1950 with stud-framing concepts.
Courtesy of Herb Miller

TIMBER FRAME

GAMBREL-ROOFED BARN

GAMBREL-ROOFED KIT BARN

timber-frame barns as large, farmers needed even more room to store the loose hay and straw at a reasonable price, and the hip part of the roof allowed more space.

This change did not happen over night, but from about 1850 to 1950, there was a big transformation from labor-intensive timber-frame joinery to mass-produced barns made from lighter, standard material.

With the adoption of standardized construction systems and lighter materials, came mail-order barns and build-your-own barn plans. In the USA *The Jamesway Farm Building Book* and *The James Way: A Book Showing How to Build and Equip a Practical Up to Date Dairy Barn* set out practical methods. Farmers could even order an engineered barn-frame kit from the catalogue of Sears, Roebuck and Company as did the Jenkins family of Chester, New Hampshire. Their 36-foot by 50-foot barn was delivered by railroad and built by local laborers in 1927. The Louden Machinery Company, from Fairfield, Iowa, began operations in 1867 by patenting a hay carrier. By the early twentieth century, it was one of the largest manufacturers of agricultural equipment. They also produced a book of barn plans and kits. They supplied blueprints and even built barns, ultimately selling more than 25,000 barns around the world. In Canada, Beatty Bros. Manufacturing Company, located in Fergus, Ontario, produced a huge array of agricultural equipment and machinery and eventually the *Beatty Bros Barn Book*. As well, the Canadian Pacific Railway's Department of Natural Resources published a barn plan book and manufactured barn kits that could be delivered to the nearest railroad station.

Today when they think of barns, most people picture the gambrel-roofed, red-painted building with white trim. These barns typically had main doors on their long side, were built with or without a basement and came with three or multiple bays. They used lightweight construction, made from "two by" (2-inch by 4-inch, 6-inch and so on) lumber. They were typically balloon-framed with truss rafters. This gave a clear span inside, the total of the barn's width and length — a huge capacity for hay. With the addition of the hayfork trolley, which came into vogue by 1900, farmers could easily load their barns up to the rafters. These multi-purpose barns did away with awkward middle posts, were cheap to build and simple to add to whenever the farmer could expand. Add poured cement floors in the stables and barnyard, and day-to-day chores became easier.

On the whole, farmers were not nostalgic about losing old barns or no longer building new ones in the traditional post-and-beam manner. No, the new buildings suited their increasingly particular needs, such as new sanitary requirements for milk production, and also made their work easier. Rapidly the gambrel barn's construction methods replaced European heavy-timber barns. Certainly there are exceptions to this, such as the Butterworks barn in Vermont, a traditional timber-frame barn built in 1982.

Variations can be found in other components of a farm operation that may or may not be incorporated within the barn, such as silos and corncribs.

Silos were introduced 100 years ago and at first were made of wood bound with iron hoops. Later, as they became very widely used and a larger capacity was needed, alternate materials became available: concrete, steel, and clay tiles. I have included just a few interesting ones because another whole book could be written only on that subject.

A local landmark for many years, this little barn with the big clock cupola near Gilford, Michigan is an eye-catcher.

I have included a few outstanding examples of corncribs, utilized to store the whole cob for livestock feed. Drying corncobs was eventually deemed less productive than putting the whole green corn plant in the silo to be used as silage. Also, dry weather after harvest was important and much of the north receives too much rain. But for a while after 1900, corncribs became a part of most farm operations; some farms continue to dry cobs over winter.

Some fantastic barn-building experiments were carried out by industrialists with vision and plenty of money. An outstanding example was Ohio C. Barber, who planned and named the city of Barberton, Ohio, after himself. By 1920, he built an incredible complex of 102, mostly brick buildings that housed everything from pigs to dairy cows and employed more than 400 people.

A modern version of a barn, with pre-made metal trusses and many windows for natural light. There are still historic touches such as the cupola and a natural stone foundation, south of Guelph, Ontario, near the village of Morriston.

The next wave of construction brought the specialized concrete and steel barns which evolved after the Second World War. These were long, low, one-story stud-framed or pole buildings, with manufactured trusses, and metal roofs and walls. These are the most common agricultural buildings today, well suited to current intensive production methods.

In an odd way agriculture has returned to its past, for in traditional English methods of agriculture of the sixteenth and seventeenth centuries, farmers housed each type of livestock and crop in separate buildings.

Today, more than one-half of the barns built before 1920 still stand. Farmers have adapted, melding new concrete and metal buildings into tall post and beam barns. But, slowly and surely, more and more old barns are unused, unmaintained, and left to the weather, falling down eventually or being demolished as a tax burden or eyesore.

Our old barns are rich in history: the history of rural life and agriculture, and the history of building by craftsmen. They also figure largely in our cultural heritage, so completely based on agriculture until very recently. Fortunately, many barn preservation societies have formed in the last twenty years, working to preserve these outstanding barns and recognize them as part of our heritage.

Standing near St. Paris, Ohio, this barn and house embody the builder's stately vision. Columns on the house reinforce a sense of grandeur. A late nineteenth-century Victorian barn, this hip-roofed structure would be ordinary without the additions of the cupolas and dormers, adornments only a gentleman farmer could afford.

Still one of the largest Shaker barns left in North America, this barn was on the cutting edge of barn building when it was constructed in 1854.

Enfield Shaker Village

The Shaker sect seceded from the English Quakers when they moved to America in 1747 and lived in agricultural communities until the mid-twentieth century, when there were only a few adherents left. They received their name because of their characteristic trembling when filled with religious ecstasy.

The Enfield Shakers in New Hampshire, named for the small town nearby, farmed at this location from 1793 until 1923. The first Shaker missionaries arrived in this area in 1782, an offshoot of the parent United Society of Believers in the Second Appearance of Christ at Mount Lebanon, New York. They originally moved to separate farms but these were difficult to adapt to their communal lifestyle. This, combined with hostility toward their religious beliefs, led these Shakers to buy a large tract of land near Enfield. Here they were able to live communally and built a village that lasted more than a century. At its heart is the largest building ever built by Shakers, the Great Stone Dwelling. The six-story granite building was finished in 1841. A beautiful mill, built nearby of the same massive stones, was completed in 1849.

Still one of the largest Shaker barns left in North America, the barn called West Meadow was heralded at the time of its construction as one of the most expensively built barns in the USA, costing $15,000. The timber-frame structure is 200 feet long and 50 feet wide. Even more interesting are its many innovations, impressive for the era in which it was built. Both main ramps into the barn are on the gable ends, so that

A variation of the bank barn, this 1854 Shaker barn has a ramp at both gable ends enabling horse and wagon to pull in, unload, and exit without turning around.

The 200-foot Shaker barn was a well-designed modern building in its time.

a wagon pulled by horses could unload the hay or other crop at any place in the barn and then come out the other end without having to turn around. It was built to house 50 milking cows, whose stalls were set on the south side to gain heat from the winter's sun while the hay was stored on the north side of the barn to keep out wind and cold. The ground floor was made of planks on a base of clay, much like mortar. It was easier to clean up manure from a smooth wooden floor than from uneven ground.

Strolling through the historic Enfield Shaker Village I instantly felt the serenity of the place. Little pathways connect the various houses and outbuildings, which are now a living museum, with the Great Stone Dwelling serving as an inn. A friendly woman in Shaker period clothing greeted me in the main museum building, which houses the simple furniture of the Shakers. She explained in a quiet voice how the barn once served as the heart of the village, providing food, gratifying work, and a good life.

Pennsylvania Star Barn

I was driving on Interstate 283, near Harrisburg, Pennsylvania, trying to find the Star Barn when I spied a steeple rising out of a new housing project beside the highway. I thought it was a church but as I sped by and looked down, I recognized the Star Barn, with its white cupola and weathervane sticking above the highway. It all happened in a flash and it took me moments to realize I could not stop on this busy road but had to keep going and find the nearest exit ramp to reach this prize barn.

I was not disappointed in the barn, but the wonderful buildings are stuck down a dead-end road in the middle of a new housing complex and their future looks bleak. The Friends of the Star Barn are trying to save this landmark and three acres left around it. If they fail it may go the way of the wrecker's hammer.

The incredible 1872 Star Barn, one of fifteen Gothic Revival barns built in the Harrisburgh, Pennsylvania, region is the only one left. The barn was constructed to showcase horses for owner John Motter, a gentleman farmer and banker.

The Star Barn and its outbuildings were built in 1872 by famous master carpenter and designer Daniel Reichert, for gentleman farmer and banker John Motter. A Gothic Revival barn, it is the last monumental barn built of the 15 originally constructed in three counties.

What strikes visitors at first glance are the large stars repeated at the top of each of the four sides of the barn, and all the other outbuildings, which, besides their decorative attraction, function as ventilators. In folklore, the five-pointed star is a protection against demons, the six gives a perfect marriage and eight provides you with perseverance!

The barn is larger than most in Central Pennsylvania, measuring 110 feet long and 75 feet wide. To the gables it reaches 40 feet and an amazing 65 feet to the top of the cupola. The barn

This is the ornate carriage house built for John Motter's fleet of buggies and wagons.

has three threshing floors, rather than the usual one or two, plus a third floor where hay or grain sheaves could be stored.

John Motter meant the whole barn, but especially the first floor, to showcase his valuable horses. This floor is higher than usual to accommodate the prize horses and there are beautiful details inside such as cross paneling of the Dutch doors and chamfering of the floor joists and ventilator posts.

Under the ramp to the threshing floors is a stone-vaulted root cellar and, at the very top of the cupola, there was at one time a weather vane punched with hearts and arrows, inscribed with Motter's name and the date of construction.

Although this will never be a farm again, let's hope it will be preserved to house a museum highlighting the agricultural history of the area.

The gothic-style vents and star, also used as a vent, were signature designs of famous master carpenter and designer Daniel Reichert.

Hard to believe this is actually the pig barn! The amazing cupola has small outlets at the very top, beneficial for birds and air movement.

New York Zenda Farm

Two fascinating pieces of history attracted me to this farm near Clayton, New York. One is the connection with Shakespeare that determined how the farm got its name; the second is the number of Jamesway steel-barn kits and equipment built on this farm during the 1930s that you can still see.

In 1915, Shakespearean actor James Hackett purchased the farm from a New York City man. Hackett was enthralled with the beautiful Queen Anne-style house on the property and the incredible scenery of the St. Lawrence River and Thousand Islands. From this playground of the rich, daily trains ran directly to the great metropolis.

James had followed in the footsteps of his father, also an actor, who had appeared on the London stage in 1828 as Richard III. Although James graduated from Columbia University Law School, the lure of the stage won him over. His earliest and most successful role was in *Prisoner of Zenda* by Anthony Hope. When Hackett bought the farm on the St. Lawrence he named it after his most triumphant play.

State-of-the-art farming triumphed when Merle Youngs bought the farm in the 1930s. He kept the name Zenda Farms, transforming the property with the help of Jamesway barn kits and equipment. The James Manufacturing Company from Wisconsin supplied all the equipment needed for a modern farm operation of its day, including barn kits that could be put together on site.

Merle built eight of the shiny Jamesway barns into an efficient mixed-farm operation using some modern agricultural methods that remain in use today.

The revamped Zenda barns and storage facilities were built during the 1930s from Jamesway barn kits.

Zenda Farm as it looks today, near Clayton, New York.

The Zenda farm during the 1950s when it was producing and bottling fine quality milk, cream and butter in the state-of-the-art Jamesway steel buildings.
Courtesy of the 1000 Islands Land Trust

When the farm was donated to Thousand Islands Land Trust, the former showplace was rusting into the ground and subsequently the non-profit organization raised funds to paint the eight buildings and storage facilities to resemble the original patina. Courtesy of the 1000 Islands Land Trust

A showplace farm of Jefferson County, Zenda Farms had a creamery with the first automatic bottler in the area. Youngs' "Golden Guernsey Milk and Cream," with a picture of a coffee maker on its bottles, was delivered to local towns for 18 cents a quart. The dairy barns had electric fly-killing screens. Every morning at milking time, workers washed each cow's tail in a special solution for hygiene. The Golden Guernsey milk was bottled in the spotless creamery and stored in a walk-in cooler until delivery time. The steel silos held corn and chopped hay; molasses added to the cattle feed enriched the cows' diet.

After Merle died in 1958, the farm traded hands. In 1991 the beautiful house on the banks of the St. Lawrence River burned to the ground.

John and Lois MacFarlane donated Zenda Farms in 1997 to The 1000 Islands Land Trust (TILT), formed in 1985 to preserve and protect the natural and historic splendor of the region. TILT undertook, through fundraising efforts, to bring Zenda's Jamesway barns back to their original state. The many activities of The Zenda Trust include art shows, a petting zoo, stargazing, a dog park and, of course, an agriculture-themed historic facility.

This facility had signs for every building, in keeping up with new scientific methods of farming.

One of the shiplap-clad walls of the barn where customers pass to the nursery business. Some of the massive posts of this 1889 barn are 25 feet long to the purlins. On the left side of the second floor were stanchions where milking cows were originally kept.

Vermont Nursery Barn

Some dreams do come true! Don and Lela Avery bought a ramshackle farm about 25 years ago with the vision of creating a garden and nursery business in this north central part of Vermont, near Morrisville. The sun peeked through the clouds as I enjoyed driving the quiet little dirt road to this farm. The route follows a small river that winds around the large hills and ends peacefully at the Avery Nurseries.

The Averys told me when they took over the property, the house was in desperate need of repair, an old asphalt tennis court covered the yard, neglected fields were full of junked machinery and the barn full of old hay, straw, and trash. Today the house has been beautifully upgraded, the yards are full of sweet-scented flowers and shrubbery and the barn centers a thriving garden and nursery business.

The three-story barn is the kind that makes you shake your head with admiration at its high drive to the third floor. Built in 1889 after the original barn burned, its seven-bent, 80-foot frame has some ingenious features, thanks to the dairy farmer who built it. Jersey cows were kept on the second floor because the farmer thought it would be warmer and that they would benefit from the natural light. For manure removal he used a clever method, up-to-date for its time. Behind the Jerseys stretched a long, narrow wooden trap door in the floor. Manure shoveled through the hole landed in a litter carrier, which then could be dumped outside on a pile. Each cow, facing inward, had the latest tie-up stanchion. The water bowls were filled from a windmill-driven pump. Hay and straw, kept on the second and third floors, could be easily pitched down for the cows' feed and bedding. The barn was so large that all the machinery and equipment could be stored out of the weather on the ground floor.

Interestingly, parts of the frame use timbers from another barn, perhaps even posts and girts salvaged from this barn's predecessor. The frame has an unusual spruce timber formation, where the ground-floor posts, which are 10 feet high, hold up the other two floors above, as in a stud-frame house. Starting from the second floor are 25-foot massive posts that go up all the way to the purlins and support the third floor and roof.

When the Averys bought the farm in the late 1970s, there was an unused asphalt tennis court in the back yard, a house in much need of repair and a barn full of old hay and junk.

Twenty-five years later, the grounds in front of the barn have been transformed into a thriving nursery business.

A great example of a two-story timber-frame corncrib with slatted walls to allow for air circulation inside. It was built on stone pillars for further air circulation and to keep rodents out.

New York Corncrib

Here is a superb example of a corncrib, built from local pine timbers in the late nineteenth century. The two-floor timber-frame structure was built to store and dry cobs of corn used to feed livestock, a source of high energy for the cold snowy months of winter in these rolling hills of Oxford, New York.

The two long side walls are slatted to allow air circulation to keep the corn dry, minimizing spoilage. They also slant inward to keep rain and snow from entering through the slats. Two slatted walls the length of the inside, about 4 feet in from the outside walls, create two long bins on either side of a middle gangway. Typically, at harvest, whole corn plants were tied in bundles, put on a wagon and then pitched into the corncrib through a square opening at the gable end. The corn plants were then untied, the cobs shucked and each cob dropped into the bins of the crib to dry naturally.

The foundation is unique for a corncrib, being 12 stone pillars which elevate the structure about two feet off the ground to help keep rodents out. The slatted floor further allows for air circulation and drying.

Corncribs were once plentiful throughout the countryside when corn became a major farm crop and are still used, although newer ones are usually made of steel mesh.

Elegant hand-forged wrought iron hinges on the door of the nineteenth-century corncrib.

Ohio Barberton Farm

This amazing 230-foot building, known as the Brooder barn, was built around 1900 as part of the Anna Dean Farm, which housed at one time the largest chicken incubator in the world. It was home to 30,000 White Leghorn chickens that were put out every day to eat insects in the nearby fields. Total production was 6,000 to 12,000 chicks a day incubated in the basement of the brooder barn.

This set of extraordinary castle-like barns was built at the beginning of the 1900s by an industrialist who planned the town of Barberton, Ohio, just south of Akron, around them. There were originally 102 buildings built on 3,500 acres by Ohio C. Barber, who called the farm Anna Dean, after his daughter and son-in-law.

Barber was a self-made millionaire, who went from selling homemade matches to heading the Diamond Match Company, the American Strawboard Co., National Sewer Pipe, and the Stirling Boiler Co. After founding the city of Barberton in 1890 he continued to expand his business and had cornered the world market as a manufacturer of matches by the time he was 60 in 1901. The farm was an experimental agricultural business that employed up to 400 workers from 1909 and ran until Barber's death in 1920.

This industrialist also believed that farming was the backbone of America's prosperity and power and that all farm buildings should be pleasing to the eye and built to last many lifetimes. Barber believed that by rejuvenating the soil and following sound husbandry practices, a farm could be operated as efficiently and profitably as any large industry. It is not surprising that he said, "The Anna Dean farm can give you what you need, all that you desire, and more than you deserve."

The barns are built in the French Renaissance Revival style of architecture featuring red brick, white concrete block, red terra-cotta tile roofs and deep blue trim. Of the original 30 magnificent brick barns, eight remain. Three of those are in the good hands of the Barberton Historical Society. This active group, founded in the 1970s around the issue of preserving the Anna Dean farm buildings, was instrumental in saving these last eight.

Besides the main livestock, which included milk cows, beef, Belgian work horses, pigs, and chickens, the farm also raised turkeys, pigeons, sheep, ducks, guinea hens, and dogs. Other products were up to 200 barrels of flour per day and 6,000 pounds of honey annually.

"The Pork Palace", as it was called by locals while in operation between 1912 and 1915, was part of the Anna Dean Farm. Ohio C. Barber selected two hardy breeds for what he named as the Piggery: Chester and Berkshire White, which he thought would be well adapted to Ohio's four demanding seasons. But in 1915 a case of Cholera was found in the swine, the pigs were moved to another location and the building housed sheep.

When the 300-foot red, white and blue Piggery was built in 1912, it cost O.C. Barber $50,000 (about $750,000 in today's dollars). The boss and his family lived in the middle upper section, which looked down on the two wings of the Piggery, later renamed the calf barn. As in all the Anna Dean buildings, the Piggery was always kept clean and fresh.
Courtesy of the Barberton Historical Society of Ohio

The farm had the largest set of greenhouses in the world, covering a total of 12 acres. Each season, it took 100 carloads of coal to heat the large greenhouse complex which was among the buildings heated and powered by four Stirling boilers. The greenhouses produced an abundance of vegetables, flowers, and fruit including 1.2 million cucumbers annually, 3,000 roses in one day's cutting and such exotic fruits as figs and nectarines.

The top of Buck's Hill makes a dramatic setting for Butterworks Farm, where both the granary and barn were built with timber-framing methods in 1983. Today the beautiful barn and granary are the hub of Butterworks Yogurt production.

Vermont Butterworks Farm

There's a little winding road near Westfield, Vermont, that climbs the mountain for about a mile before it levels off. Another four miles takes you to the top of Buck Hill, a spectacular view, and Butterworks Farm. This organic dairy farm began in 1975 with three Jersey cows and has grown into the number one yogurt producer in Vermont. Committed to organic agriculture, owners Jack and Anne Lazor grow all the feed needed for their Jersey herd, which gives them the milk for their high-quality yogurt.

Their barn and three-story granary could be fine examples of traditional timber framing of the 1800s, but in fact were built in the early 1980s. The timbers were milled locally and a Vermont timber framer chiseled the traditional joinery and raised the barn with the help of many people. Just like an old-fashioned raising bee!

The barn stands beside the little municipal dirt road, which dead-ended at the next door neighbor's. With a high drive to the third floor, and more and more farm machinery, employees' cars, and tractors parked around the barn as business grew, the local municipality decided to build another road around busy Butterworks Farm.

Although the barn is used for such conventional farm work as milking cows and storing hay, it also houses a "little factory" on the first floor where the Jersey milk from the stables downstairs gets converted into rich yogurt.

I visited Jack and Anne in the summer and one evening after milking, I helped Jack take the cows out to their nightly pasture. It was late in the season and on top of this mountain, the air was fragrant from fresh-cut sweet hay, pungent oats and sunflowers, and the stars above shone like so many crystals in the sky.

The barn frame goes up. Barn raisings were a community effort during the 1800s, when most of them were built. At Butterworks Farm this tradition continued, but with a few more modern aids. Courtesy of Jack Lazor

The interior of the Butterworks barn built with traditional timber-framing methods in the 1980s.

The posts and beams of this late nineteenth-century barn were notched by a master framer with the help of his apprentices. Typically he would make a list of timbers needed for the barn and have a crew cut the appropriate trees during the winter. They would be squared while green, with adzes and axes and then transported to the barn site. There the master framer would draw each notch on the timbers and his apprentices would then cut the notch with boring machines, saws, and chisels. Once everything was finished, the neighbourhood would be invited for the raising bee and the heavy bents would be lifted up by many men and then pushed into place using pikes and poles. After the raising, the women would put on a great meal for everyone involved and a dance inside the barn would follow. (See barn raising photo on page 13)

Silos come in all shapes and sizes, but this narrow, all-cement silo is exceptionally pillar-shaped. Standing 45 feet tall near Palmyra, Michigan, it has a molded top featuring clean, elegant lines.

An astonishing sequence of roofs which somehow echoes the landscape in which it is set. The house is connected to the English barn, which is connected to the wooden silo, which is connected to the small shed, which is connected to another barn, which in turn is finally connected to another shed. It's a great example of connected barns, near Ludlow, Vermont.

Many barns were used for different purposes. The Walker Tavern Barn, near Cambridge Junction, Michigan, built in 1836, was originally used as a barn, a blacksmith shop and a tavern. As travelers passed by in their horses and buggies, the need was great to refresh oneself, since being on the road was slow and sometimes muddy or dusty. Farmers along more traveled dirt roads often opened their houses as inns and their barns as drinking places.

Old barns are now often adapted for other purposes such as this antiques business in Milford, New York. Built in 1900, the gambrel-roofed barn has four floors of antiques and curios, with the post and beam structure complimenting artifacts of the same era.

Is it a barn or a castle? This question must run through the minds of thousands of visitors who tour Shelburne Farms on Lake Champlain every year. William Seward and Lila Vanderbilt dreamed of a grand model of agriculture when in 1886 they bought the 3,800 rolling acres which became Shelburne Farms. The grounds were designed by world-renowned landscape architect Frederick Law Olmstead, who also designed Central Park in New York.

Robert H. Richardson, one of the most prominent architects of his day, designed all the handsome buildings, including five barns. His intention was to create the feeling of an English country estate, with round turrets, gabled roofs and Tudor detailing.

Today the 1,400-acre farm is a National Historic site, where, since 1952, a herd of Brown Swiss cows have provided milk, the basis of the cheddar cheese Shelburne Farms is known for across the USA.

These incredible buildings are part of the Farmers' Museum in Cooperstown, New York. The museum lands have been part of a working farm since 1813, when the family of novelist and adventurer James Fenimore Cooper owned them. His father, who was a US Congressman, moved to upstate New York and founded the town Cooperstown when James was a one-year-old. By 1918, a modern dairy operation was operating at Fenimore Farm. At that time the main barn was built from local cut stone in the Colonial Revival style. In 1944 the farm opened as the Farmers' Museum and today three of the original buildings still stand: the barn, the herdsman's cottage and the creamery. A whole farm and pioneer village have also been recreated for visitors to be able to see farm life of the 1800s.

A view of Ridgeview stables in Morriston, Ontario, through the old stone barn foundation, shows the garden area.

Ridgeview's modern horse barn centers a 148-acre horse paradise which includes an 80-foot by 180-foot indoor arena, a 100-foot by 215-foot outdoor arena, a heated viewing area, eight lush paddocks, complete fire and alarm system, ten miles of horse trails and a gated entrance. The arena, stables, and entrance room were built in 2002 at a cost of $3 million. Fine stonework finishes off the exposed part of the foundation all the way around the arena complex and many choice details grace the barn's interior.

A metal truss system supports the metal clad roof, with a stud framing combination for the walls. The stables are made from different types of hardwood, such as oak and white ash and metal grating for the horse's safety, while the visitor's area is well appointed, including exposed wooden trusses.

Horses are most healthy when they are outside as much as possible. "They need to mimic nature," explained Doug Dean of Durham Wood Farm, a tall, craggy-faced man with a passion for his work. Doug should know. He has been in the horse breeding business since 1981, has been recognized as the Leading Breeder of Event Horses in America and has, on average 50 sport horses on his 50-acre farm, near Durham, Ontario.

When Doug Dean bought the farm there was an old house and an 1880s bank barn on the property. Today, there is a 60-foot by 140-foot riding arena, a hay and loafing barn, stables added to the bank barn and another addition to the arena plus a partly covered straw paddock area for the horses to lie in. With all that roofed area he still leaves it up to his horses whether to stay outside or not. "Most often," he explains, "they'll stay out. They know what's best for their health."

The combination of new buildings, built with manufactured truss roofs and pine barn board and the historic barn has worked well for him. In the old haymow, Doug still stores small square bales of hay which he feeds to his horses through a trap door to the renovated stables below. In the hay/loafing barn, horses feed on large square bales, but they are not restricted in their movements, since the building is open on the south side. The horses will come into the barn on cold winter nights or during heavy rain and wind or ice storms.

Always a wondrous sight, the inside of a post and beam barn. This one is of the queen-post type. The ladder was used to reach the top of the peak to fix and adjust the hayfork, still hanging from the ceiling.

Built mostly of oak, the 38 box stalls usually hold new foals and unpredictable stallions used for breeding. This hard-working horseman runs the operation by himself, with occasional help to exercise the horses. "You can breed good horses, but you need good riders for them to excel so it's all luck," says Doug with a smile.

Set in a beautiful valley near Craftsbury Common, Vermont, these two connected timber-frame barns measure more than 150 feet long with a dramatic covered high drive to the third floor, just like some of the round barns. The gambrel roof dates it to the late nineteenth century.

At one time the road went under the amazing covered high drive of this double barn, near Craftsbury, Vermont. Covered barn bridges are usually found at round barns, where the horse and wagon can enter and go around the circular floor inside and then leave without having to turn around.

Looking like a long barn, and built with some of the same timber-framing techniques, this is the Cornish-Windsor Bridge. At 460 feet it is the longest covered bridge in the USA, spanning the Connecticut River between Vermont and New Hampshire. Using the patented Town Lattice Truss system, the all-wooden structure is notched and pinned using traditional joinery methods and bolted for extra strength. Built in 1846, it was the fourth bridge on this site. The other three, beginning with the first in 1796, were washed away by floods.

Two covered entrances lead into this gambrel-roofed barn, recently painted and standing dramatically in the hills of Sutton, Quebec.

Today's dairy farms still try to emulate the look of the old, with red roofs, plenty of vents and fancy dormers. The newer buildings, such as this gambrel-roofed barn, are much lower because they don't have to accommodate the huge amounts of loose hay early timber-frame barns had to store. The round bales now mainly used are stored in separate buildings. But the new barns are longer, to accommodate the larger herds needed to make a living today. This stud-framed structure is located on the Lake Champlain flats, near Charlotte, Vermont.

A pleasing limestone building built in the once-prosperous area of East Springville, New York. Initially this 1830s structure was constructed for the Francis Carriage Works. Thomas Francis built some of the finest carriages and wagons in the area at a time when communities were more self-sufficient and just about everything needed was built or grown locally. During this period, nearby farmland was used to grow hops for the production of beer. Many fine houses and barns were constructed with the capital garnered from that lucrative crop.

Francis Carriage Works was later used as a cabinet-making shop and then a barn to store grain and hay. During the 1776 Revolutionary War, English soldiers, known as the King's Rangers, raided East Springfield and massacred the residents at nearby Cherry Hill. Today, in the fields around East Springville, people still find bits of history, such as broken swords and buckles lost when the English and Revolutionary armies clashed in this region.

Wondrous timber-frame barns, such as Doug Dean's near Durham, Ontario, still have many uses: first as part of modern farming enterprises for storage and equipment, or as recreational property, as a perfect place to play and dance in, or even as a home. These amazing buildings will never again be built, so it's important to preserve them as part of our common history. Courtesy of Doug Dean

Dictionary of Terms

Bent: A series of timbers that makes a structural frame with a minimum of two posts and one horizontal girt.

Brace: a short piece of small dimensional timber connecting an upper and a lower timber at an angle of 45 degrees.

Connecting girts: horizontal beams that join bents of a barn.

Collar tie: a horizontal beam used between two rafters to support the roof.

Dormer: a projected opening built out from a slope in the roof.

Embrasure: a slit in a stone barn wall.

Forebay: also called an overshoot or laube is an overhanging mow that protects the barn wall below and the animals in the barnyard.

Gable: the triangular upper section of the end wall under two sloping roofs.

Gambrel roof (also called **hip roof**): a roof with its slope broken, the lower section being a steeper pitch than the upper.

Granary: a wooden storeroom inside the barn for threshed grain.

Gun stock post: a post that is wider at the top than the bottom to add more wood for joinery.

Hayfork: a mechanical device used for lifting hay from wagons to a hayloft or mow.

Hay hood: a small roof on the gable end of a barn protecting an outside hayfork.

High drive: a ramp to the second story of a barn.

Jerkinhead: a clipped gable.

Litter carrier: a rectangular box used to transport animal manure and used bedding out of the stables of the barn. Also called a manure caddy.

King post: one or two center posts that extend to the purlins or collar tie.

Mortise: a rectangular slot drilled and then chiseled into one timber to receive the tenon from another.

Mow: the place in a barn where hay, straw, or grain is stored.

Post: a vertical timber.

Post and beam: another term for timber-frame construction.

Purlins: within the roof structure, horizontal timbers that support the rafters.

Queen post: two vertical posts extending from the top plate or girt supporting the roof.

Rafter: one of the main poles or beams supporting a roof.

Saltbox: type of structure with one end of the roof being lower than the other.

Scarf: a notch for splicing two large timbers end to end.

Shakes (also called **shingles**): thin wedge-shaped material, historically cedar used with others to cover a roof or side of building.

Sill plate: horizontal timber on top of the foundation.

Sleeper: a horizontal log, usually with only the top flattened, which supports the barn floor.

Stable: lowest part of the barn where livestock are kept.

Stook: bundle of sheaves of cut grain, arranged stem-end down to dry.

Stud framed: using milled two-inch wood to build a structure.

Tenon: the tongue chiseled at the end of a timber so that it will fit into the rectangular hole of the mortise.

Threshing floor: an area on the main floor of the barn where crops are hand- or machine- threshed.

Timber frame: a structure made from large timbers, joined and pinned together supporting smaller timbers on which the walls, roof, and floor are affixed. Same as post and beam.

Top plate: horizontal timbers that support the bottom of the rafters.

Transom: a lintel above a door.

Bibliography

Arthur, Eric, and Dudley Witney. ***The Barn: A Vanishing Landmark in North America.*** Toronto and Greenwich, Connecticut. 1972.

Benson, Tedd, with James Gruber. ***Building the Timber Frame House: The Revival of a Forgotten Art.*** New York. 1980.

Boyce, Gerald E. ***Historic Hastings.*** Belleville. 1967.

Bread and Puppet Museum (brochure). Bread & Puppet Museum. n.p. n.d.

Cazaly, Peter D. ***Schoharie Barns.*** n.p. n.d.

Cuff, David J. et al. ***Atlas of Pennsylvania.*** Philadelphia. 1989

Cumberland County 250th Anniversary Committee. ***Pictorial History of Big Spring Area.*** Carlisle, Pennsylvania. 2000

Davidson, Stanley E. and May Davidson. ***Londonderry and Skull Fork Valley Memories.*** n.p. 1981

Fitchen, John. ***The New World Dutch Barn: a study of its characteristics, its structural system, and its probable erectional procedures.*** Syracuse. 1968.

Gunning, Margaret W. ***The History of Freeport, Ohio From 1810 to 1900.*** n.p. 1991.

Hainstock, Bob. ***Barns of Western Canada: An Illustrated Century.*** Victoria, BC. 1985.

Île d'Orléans Tourist Guide. St. Pierre-de-l'île-d'Orléans. 2005.

Kauffmann, Carl. ***Logging Days in Blind River.*** Sault Ste. Marie, Ontario. 1970.

Klakin, Charles. ***Barns: Their History, Preservation and Restoration.*** New York. 1979.

MacDonald, J. E. ***Shantymen and Sodbusters: an account of logging and settlement in Kirkwood Township, 1869-1928.*** Thessalon, Ontario. 1966.

Rawson, Richard. ***Old Barn Plans.*** New York. 1979.

Rempel, John I. ***Building With Wood and other aspects of nineteenth-century building in central Canada.*** Toronto. 1967.

Ritchie, T. et al. ***Canada Builds 1867–1967.*** Toronto. 1967

Smith, Elmer L. ***Hex Signs and other Barn Decorations.*** Photographs by Mel Horst. Witmer, Pennsylvania. 1965.

Sullivan Historical Society. ***A History of Sullivan Township 1850–1975.*** Desboro, Ontario. 1975.

Van Dolsen, Nancy. ***Cumberland County: An Architectural Survey.*** Carlisle, Pennsylvania. 1990.

Whitton, Charlotte. ***A Hundred Years A-Fellin', 1842-1942, Some passages from the timber saga of the Ottawa in the century in which the Gillies have been cutting in the Valley.*** Braeside, Ontario. c1943, repr. 1974.

WEB SITES

Barn Again **www.agriculture.com**

Barn Journal **www.thebarnjournal.org**

Hudson River Valley Review **www.hudsonrivervalley.net/hrur/about.php**